The Life Pill

The Life Pill

Why Not Take Life for Life?

Alfred Sparman, MD

THE LIFE PILL
WHY NOT TAKE LIFE FOR LIFE?

iUniverse books may be ordered through booksellers or by contacting:

iUniverse
1663 Liberty Drive
Bloomington, IN 47403
www.iuniverse.com
1-800-Authors (1-800-288-4677)

ISBN: 978-1-4917-8403-7 (sc)
ISBN: 978-1-4917-8404-4 (hc)
ISBN: 978-1-4917-8405-1 (e)

Library of Congress Control Number: 2015920070

Print information available on the last page.

iUniverse rev. date: 02/24/2016

To my mother, Olga Doris Sparman.

A special thank you to Lena Wills, Dr. Nicole Moore-Clarke, Annatasha Sparman, Kimberlee Thompson, and Stephanie Sparman for their endless support in this breakthrough.

Also special thanks to the Sparman Clinic and the 4H Hospital.

Reach for the stars, and if you make the treetop, you're still above ground level.

—Alfred Sparman, MD

Other Works by Alfred Sparman, MD

1.58 Seconds

Switched

Contents

They say be careful what you wish for; you may get it … I wish for a long life.

—Alfred Sparman, MD

Introduction

My boyhood years were kind of intense. Out of a class of thirty-two students, I would say a third of the class had a desire to follow the pathway of science. I still remember the days when I would dissect frogs with a knife. I just wanted to see what was happening in there. I thought it was kind of strange at that point, because most boys seven to eight had toys of a different nature. But as a child, life offered me thousands of questions, and there seemed to be few answers. I was always one of the hardworking, bright kids in class—not because I was a genius of any sort but because I tended to place a lot of energy and focus on any engagement even from a very early age. My mom was a nurse and my dad a police officer, which explains why the medical field and law enforcement were some of my choices in the early years. Mom and Dad worked hard to make the best out of their eight children, of whom I was the third. We lived in a two-bedroom house, and the rest you could figure out.

Science came pretty easy to me; the formulas and laws of early scientists were things that I found extremely interesting as opposed to history and literature, which seemed to be beyond my reach. I remember once, while attending Long Island University, I received a C, which would have been one of the few Cs I received during my college years. I was so devastated that I went and purchased a tape recorder to tape the lectures. I figured if I could tape everything the teacher said, I would at least get a B. I remember once I entered the class and the professor said, "Mr. Sparman, I pity you having to listen to my voice twice." I still ended up getting a C in that class.

My mom and dad were very religious, and we as a family were always churchgoers. I was extremely zealous while in the church. I remember many times going with my brother to the street corner and preaching or going out to various companies to give out religious materials. I was preoccupied from an early age with the living and dying process—what did it mean? How could one person live for just one year, another one day, another seventy years, and another a hundred years, and yet the world continued as if it did not care? Then would I reflect on my first victory: out of a hundred million sperms, I'd won. That was a miracle! With that concept, there came a sense of purpose, and with that purpose came a fulfillment of that purpose; that was my thought process even as a teenager.

Frankie, a very close friend of mine, died by drowning at an early age. I could not understand that. I wondered if I could have done something to save him or help him. I wondered if it was his end or his beginning. He'd been a good kid and extremely friendly; these were my thoughts as a young teenager. I believed then—and I believe now and will always believe—that everyone comes here to make a contribution on this planet and that our duty as human beings is to find our callings and leave something behind that will be beneficial to our fellow human beings.

I received a scholarship from Long Island University to study medicine at New York Medical College, and the questions continued to escalate. I can assure you that medical school answered some of them, so I was encouraged to continue along that path. St. Luke's Roosevelt Hospital in New York City was where I did my internship and residency in internal medicine. Urology was my first choice, but I think the reason I eventually chose cardiology was that it was the number one killer—and it still is. *Maybe,* I thought to myself, *the answer to longevity is somewhere within cardiovascular diseases.* I have a very strong family history of cardiovascular disorders. So the journey to remove some of these questions from my thoughts had begun. I thought of Newton, Einstein, Galileo, and others who, I presumed, must have had similar questions that caused obsession with an eventual positive outcome. So I started to look for ways in which I could link the broken chains that some of my contemporaries might have overlooked. My first interest was the human heart and how students could better define the coronary anatomy. I still remember long drives from Jacksonville, where I was doing my cardiology fellowship at the University of Florida, to Georgia to harvest human hearts from cadavers. That research did not come to fruition. As I moved on and started to practice cardiology, I observed patients who had end-stage heart disease, where medical science had reached its limit. I noticed the anxiety, reservation, fear, and confusion that embraced patients as they were about to leave to go to the other side; this reignited the fascination with life and death that had always plagued me. The life span of my mom's lineage was, on average, fifty to sixty years, so according to history, she should have succumbed to death by age sixty. However, Mum became a vegetarian at age fifty-one for reasons I cannot explain, but I think it could have been her medical knowledge and her desire to change the family formula. She lived until eighty, the longest of anyone in her family tree. Now, I was extremely close to Mom; it took me a while to find out that I was her favorite child. Her extended life span after changing her diet sparked something in me. I came to the realization that modification of one's diet could influence longevity. Well, I became the first disciple. I immediately became a vegetarian, thinking that if Mom could extend her life by thirty years using just a vegetarian lifestyle, I could take it a bit further by also incorporating other health habits, such as exercise, drinking lots of water, and so forth.

Many factors affect longevity. Studies have shown that people who are single live shorter life spans. It is also known that those who have strong social support live longer. People who pray and meditate induce endorphins and oxytocin, which are relaxing hormones that decrease stress levels. Individuals under extreme stress tend to have shorter life spans as well. Exercise has been shown to improve cardiovascular health, people who drink lots of water have healthier organs, and people who choose more plant products and less meat live healthier lives. Individuals who consume colorful plant products and antioxidants combat early death from the atomic level upward. And last but not least, people who practice appropriate sleep cycles tend to live longer.

In this book, I will build the stage for you to understand the most common causes of diseases from the atomic and cellular levels. If we understand what happens in the unit structure of the cell, we will understand what happens in the tissues, the organs, and the systems. I will take you along the path of unstable atoms and the free radical theory of aging and oxidation, which is one of the lethal reactions that induces cell death. We will understand the role of antioxidants in preventing cardiovascular disease, cancer, diabetes, arthritis, decreased blood flow, hyperlipidemia, hypertension, erectile dysfunction, inflammation, and many other noncommunicable diseases.

Bryophyllum pinnatum, or the life plant, which can survive in extremely hostile environments, is a very strong antioxidant; we will learn of its properties and the role it plays in longevity. The more commonly known *Moringa oleifera* is also a very strong antioxidant that I will describe with all of its other health-related benefits. And vitamin C, the most popular of the three, has long been known to be a strong antioxidant and an essential vitamin to combat many disease entities. The synergy of these three powerful antioxidants gives rise to something the world is waiting for—an answer, a formula, a code, a message, something to help them delay their crossover to the other side. They want something natural, something herbal, something that is not artificial or overly processed. Maybe they want something given to them like what was given to the first Adam. That something is: "The Life Pill"!

To understand science is to understand God.

—Alfred Sparman, MD

1

Aging

Today is the oldest you have been and the youngest you will ever be again.

At conception, the human body is at a primal point. It is assimilating and developing based on a multitude of factors: genetic makeup, biological factors from parents, and additional environmental stressors all play a part in mapping the overall development of our bodies. Our genes automatically create our destinies and thus guide us on paths from youth to aged versions of our former selves. In many instances, our bodies begin to give us a glimpse of our future selves around the age of thirty, when we start to notice minor glitches in what used to be well-oiled machines; our sight may waver, our joints may moan, and our ability to fight off infections may be diminished. As we grow older, we become prime candidates for a multitude of illnesses that we may not be able to ward off, because of either genetics or psychological disposition. Aging is one of the largest known risk factors for human disease, with an average of a hundred thousand people dying worldwide each day because of age-related causes. As we age, our viability decreases and our vulnerability increases.

The statistics above seem to paint a bleak picture of aging. Nonetheless, it must be noted that in the United States alone, men and women who live to ninety and over make up one of the largest growing populations. The human life span has significantly increased, as the mean life span 150 years ago was around 40 years old, while 1,000 years ago, it was around 25 years old. The oldest living person thus far, Ms. Jeanne Calment, was born in February 1875 and lived to age 122 in France. Currently, the oldest living person is Ms. Susannah Mushett Jones, who lives in Brooklyn, New York, and who celebrated her 116th birthday on July 6 this year. We are seeing more centenarians than ever before. According to the US Census Bureau, a survey in 2010 revealed that 53,364 people had hit the hundred-year mark; and in the UK, feedback from the Office for National Statistics stated that there were 13,350 centenarians in 2012—a number that almost doubled that of the 2002 report. The average life expectancy for a newborn girl in the UK in 2012 was eighty-two years, and for a boy, it was seventy-eight years (Census 2010). We have done wonders with extending life. However, some animals perfected this technique centuries before us. The Galapagos giant tortoise can live to the ripe old age of 190 years without even looking a day older! Bowhead whales have a life span of one hundred to two hundred years. Animals and humans share common factors that may prove to be detrimental to their longevity. These include our habitat or environment, our diet, and our lifestyle. We have a fair understanding of how these factors are intertwined with aging; however, the biggest challenge we face is understanding the mechanisms of aging on the cellular and molecular levels. Research into aging is extremely time consuming and costly.

Here is what we understand thus far about aging. Aging is defined as the gradual change in an organism that leads to increased risk of weakness, disease, and death. It takes place in a cell, an organ, or the total organism over the entire adult life span of any living thing. There is a decline in biological functions and in the ability to adapt to metabolic stress. For example, changes in organs include the replacement of functional cardiovascular cells with fibrous tissue. Overall effects of aging consist of reduced immunity, loss of muscle strength, decline in memory and other aspects of cognition, and loss of color in the hair and elasticity in the skin (Webster Inc. 2000). Typically, your age is measured chronologically and celebrated as a milestone. However, aging can be broken down into multiple types:

- Universal aging refers to age changes that all persons share.
- Probabilistic aging refers to changes that happen to some but not all persons as they grow older.
- Social aging includes the cultural expectations of how people should act as they grow older.
- Biological aging refers to changes in an organism's physical state as it ages.
- Proximal aging involves age-based effects that come about because of factors in the recent past.
- Distal aging refers to age-based differences that correlate to a cause early in a person's life.

Your chronological age does not always correlate with your functional age, as in many instances a person's mental maturity or physical prowess may not correlate with his or her chronological age.

Before we continue, we must note that the study of aging is called gerontology, and for many decades, researchers have been searching for the ultimate cure for this plague we call aging. Significant strides have been made, and a plethora of theories have arisen to slow down the process of aging or, in some instances, to repair the damage caused to our bodies as we age. To fully grasp the effects of aging, we have broken down the biological, physical, and societal aspects of growing older.

The Biology of Aging at the Cellular Level

At birth we begin as a single cell, a zygote, which rapidly divides to form a cluster of cells called a morula. This single cell contains our genes, twenty-three pairs of chromosomes that we acquired equally from our mother and father, along with a nucleus. Our DNA is contained within our genes and is composed of a double helix strand similar to a twisted ladder. Each step is composed of a pair of bases bonded together. These base pairs encode information, and scientists use letters of the alphabet to represent this code. The human genetic blueprint consists of approximately twenty-five thousand genes made up of approximately three billion letters or base pairs. The cells continue to multiply and move from the fallopian tube to the uterus, forming the blastocyst, which is somewhat larger than the morula. This ball of cells separates. The inner layer becomes the embryo, and the outer layer of cells is programmed to nourish and protect the embryo. As the process continues, the embryo's cells continue to multiply, with different layers of cells dedicated to creating the various organs and structures that form the human body. Major cell types include skin cells, muscle cells, neurons, blood cells, fibroblasts, stem cells, and others. Cell types differ both in appearance and function yet are genetically identical. Cells are able to be of the same genotype but different cell type due to the differential regulation of the genes they contain. Even after birth, the cells continue to divide for numerous reasons, such as to replace old, dead, or damaged cells. Most importantly, cells divide to allow us to grow; this can occur trillions of times every day. All our trillions of cells contain replicated DNA.

Genetic Imprint

Genes are made up of DNA and govern the production of proteins that form every tissue in the body. This process, known as gene expression, begins with transcription, in which a molecule called messenger RNA transfers the information in DNA out of the cell's nucleus and into the cytoplasm, where it is translated into amino acids that form proteins. These proteins are designed for specific purposes. Small deviations in the base pairs naturally occur about once in every thousand letters of the DNA code, thus creating a small genetic variant, or polymorphism. The epigenome is the chemical infrastructure that acts directly on the genes to switch them on and off. The body's complete library of DNA, known as the genome, is found in every cell. Yet only a portion of the genes within a cell are switched on to produce proteins at any given time. A variety of influences determine whether particular genes are on or off. For instance, mutations can turn off a gene or alter the types of proteins it makes. Moreover, the epigenome creates a pattern of modifications that help determine which genes are turned on and off.

By helping to regulate genes' status, the epigenome ensures that a developing liver cell doesn't try to become a hair cell or a neuron. What's more, the epigenome helps ensure that patterns of gene expression are preserved when cells divide. These variations in our genetic design predispose us to a variety of physical and biological traits, such as our hair color and affinity for particular diseases or disorders.

Researchers have been seeking to determine the genes associated with longevity. For example, studies have examined centenarians and compared them with persons with average or short life spans to infer whether a genetic variant can be found. Using another approach, the candidate gene approach, scientists look for genes in humans that serve similar functions in the body as genes already associated with aging in animal models. For example, scientists found the longevity genes involved in the insulin/IGF-1 pathway of animal models. They then looked for the comparable genes in the insulin/IGF-1 pathway in humans. Then they were able to examine humans to see if a variant of the gene was prevalent among people who lived long, healthy lives and not among people who had average life spans. The FOXO3a gene is another gene variant that was predominant among long-lived individuals suggesting a possible role with longer life span (National Institute of Aging 2011).

Another approach, the genome-wide association study (GWAS), is particularly productive in finding genes involved in diseases and conditions associated with aging. In this approach, scientists scan the entire genome looking for variants that occur more often among a group with a particular health issue or trait. In one GWAS study, National Institutes of Health–funded researchers identified genes possibly associated with high blood fat and cholesterol levels—and therefore with risk for coronary artery disease. The data analyzed were collected from Sardinians, a small genetically similar population living off the coast of Italy in the Mediterranean, and from two other international studies. The finding revealed more than twenty-five genetic variants in eighteen genes connected to cholesterol and lipid levels. Seven of the genes had not been previously connected to lipid levels, suggesting that there are possibly other pathways associated with risk for coronary artery disease. Heart disease is a major health issue facing older people. Finding a way to eliminate or lower risk for heart disease could have important ramifications for reducing disability and death from this particular age-related condition.

Scientists will continue to discover and understand the genetics of aging through the use of GWAS and candidate approaches. However, because of the enormous complexity of the genome, it seems likely that scientists will never identify just one gene that is directly implicated in health and life span. Instead, they may be able to identify several combinations of genes that affect the aging process.

What Is DNA Damage?

Our DNA suffers daily from millions of damaging events. However, our cells have powerful mechanisms that repair this damage even as we age. Unfortunately, as we age, our cells cannot repair all the damage, and thus these glitches will remain in our DNA. It is believed that the combination of this damage and the body's inability to repair itself may lead to aging. There are other types of damage that can lead to modifications in our DNA. For example, when replicating, there may be small errors in the DNA code called mutations, which are typically harmless. The more damaging changes can result from a break in the DNA strand, which is more complex to repair and could lead to more mistakes during the repair process that may shorten life span. In other instances, a cell divides, passing on its information to its two daughter cells, and with each cell division, the telomere (a stretch of DNA at each end of the chromosome that protects the protein-encoding part of the DNA) becomes shorter. When the telomere becomes too short, it is no longer able to protect the DNA, leaving the cell at risk for serious damage.

Telomere length cannot typically be restored in most cells. In instances where the length becomes too short, it triggers a response leading to one of three outcomes: (1) the cell stops replicating and turns itself off, or becomes senescent; (2) it stops replicating by dying (apoptosis); or (3) it continues to divide and becomes abnormal as well as potentially dangerous (that is, it becomes cancerous). It is valid to note that senescent cells, although turned off, continue to work on various levels. For example, they may continue to interact with other cells by sending and receiving signals. However, they are different from their original selves, as they cannot die and release molecules that lead to an increased risk for diseases like cancer.

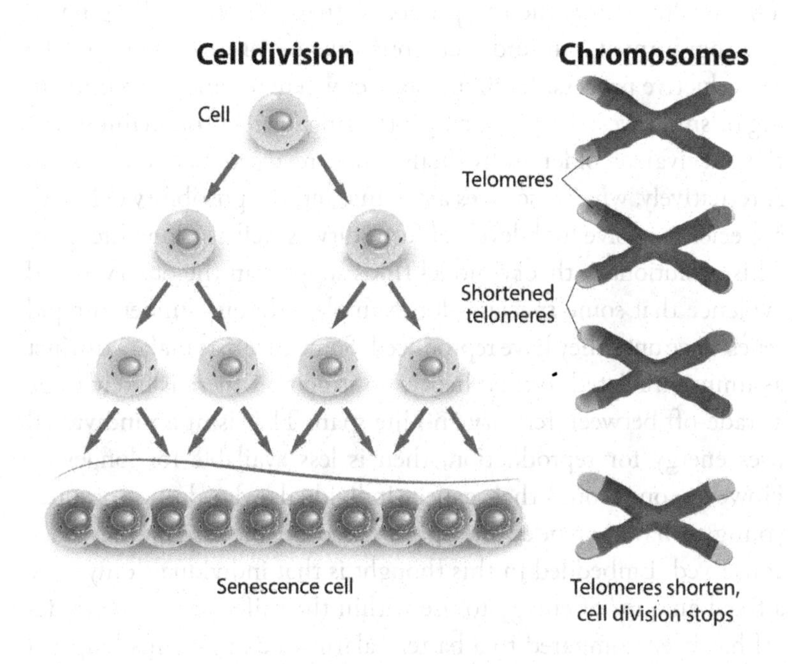

Below are some of the leading biological theories of aging (adapted from Riddle et al. 1997).

Oxidative Damage

The free radical theory of aging and the oxidative damage theory are one and the same. It is believed that aging is the result of oxidative damage. The theory proposes that all species have different life spans; therefore, long-lived organisms seem to be more resistant to oxidative damage—that is, they may have more effective ways of scavenging reactive oxygen species. These reactive oxygen species, or free radicals, are very unstable and create oxidative damage, giving rise to the symptoms we see with aging. We will examine this theory in detail in one of our later chapters.

Evolutionary Theory/Progeny Production

The evolutionary theory proposes that the rate of aging is preprogrammed, an indirect consequence of selection for the reproductive process. In circumstances where resources are limited, organisms invest in producing offspring instead of maintaining the survival of older individuals who are reproductively less fit. Alternatively, when resources are abundant, the possibility exists for a species to evolve high levels of fecundity as well as a long life span. This evolutionary theory model finds support in the observational evidence that some species—for example, salmon—undergo rapid senescence once they have reproduced. Some theorists make a distinct assumption related to this theory, positing that there is likely to be a trade-off between fertility and life span. That is, if an individual uses energy for reproduction, then is less available for longevity. However, one would then expect individuals who do not produce young to have extended life spans based on the energy they have conserved. Embedded in this thought is that individuals only have a fixed amount of energy to use within their life spans. The trade-off has been compared to a battery alarm clock that runs longer if the alarm is turned off. However, living organisms are similar to an electric alarm clock once food is not limited. Therefore, the necessity for a trade-off within an individual is somewhat unclear.

Metabolic Rate

The metabolic theory, as the name implies, states that the length of a person's life span is controlled by the rate of metabolism. The premise is that the quicker a species burns calories, the faster it will age. There exists a correlation between metabolic rate and life span in some groups of animals, such as birds and mammals with the exception of primates. However, some opposing arguments state that the metabolic rate cannot be the only determinant for the rate of senescence. For example, long-lived mutants exist in *C. elegans* that do not have lowered metabolic rates.

mTOR

There is a protein called mTOR (mechanistic target of rapamycin) that prevents autophagy, the process that manages the destruction and creation of new cells. This protein has been linked to aging via the signaling pathway. Caloric restriction in some model species has been shown to lead to a longer life span; the nutrient-sensing function of the mTOR pathway may oversee this effect. When an organism restricts its diet, mTOR activity lessens, allowing autophagy to increase. Autophagy allows for the cleanup and recycling of old and damaged cell parts, thereby increasing longevity and decreasing obesity. Additionally, researchers have linked mTOR to a reduction in glucose concentration in the blood, thus reducing insulin signaling. One theory is that the inhibition of mTOR and autophagy decrease the effects of reactive oxygen species on the body, thereby preventing damage to DNA and other organic material. This might allow for an extended life span.

Reproductive Cell Cycle

As the name states, this theory promotes the idea that aging is regulated by the reproductive hormones, which act aggressively in a pleiotropic manner—that is, one gene influences multiple genes that are seemingly unrelated phenotypic traits. In the early stage of life, this is done via cell cycle signaling that would typically promote growth and development. However, as age progresses, the body makes a futile attempt to maintain reproduction, causing development to become irregular and thus driving the cells toward senescence.

Caloric Intake

Food is an environmental influence that has been shown to have a very interesting effect on the rate at which we age. For example, rats can live up to 60 percent longer if they consume a low-calorie diet. This is quite remarkable, since calorically restricted individuals are healthy and robust, with normal metabolic rates. The only catch is that these individuals do not reproduce.

Should food be restored, they are able to become fertile again, even if they have reached the age at which well-fed rats would be past their reproductive prime. In theory, restricting calories could act as a reset button, restarting the normal mechanism that controls the rate of aging. Since different species carry different aging clocks, the possibility exists that the clock could be reset in an individual. On the other hand, restricting caloric intake could induce a type of antiaging process that is essentially different from normal aging.

The Link between Evolution and Aging

Evolution is the overarching theory that unites biology, and as a result we need to understand how it interacts with the aging process. It is thought that aging evolves because of the interaction of two effects. First, Darwin's theory of natural selection acts more powerfully on the young than on the old; this explains why the autosomal dominant diseases, such as Huntington's disease, can persist even though they are inescapably lethal. Second, any genetic, developmental, or physiological traits that benefit early survival and reproduction will be selected for even if they contribute to an earlier death. Evolution is focused on ensuring the continuation of a species and not on aging. Such genetic effects are called antagonistic pleiotropy. "Antagonistic pleiotrpy refers to genes beneficial to and even critically necessary for growth and reproduction that 'backfire' in older animals and contribute to aging, in part through 'unexpected' interactions" (Watt 2014). Genetic pleiotropy refers to genes that have multiple effects. There are genetic variants that increase fertility in youth, but as we become old, they are known to increase cancer risk. Two examples are the genes p53 and BRCA1.

Conversely, the same mutations, if they are expressed only at old age, are neutral to selection because their bearers have already transmitted their genes to the next generation. Note that these mutations may or may not affect fitness.

In addition to these biological theories, there is the maintenance theory. According to this theory, researchers and biologists should focus on maintenance medicine. The body was designed to handle damage—just not an infinite amount of it. There are seven major types of damage that occur within the cellular structure of the body, thus leading to aging. Aubrey de Grey, in a presentation titled "Seeking Immortality" at TEDxSalford 2014, listed the types of damage as follows:

1. Cell loss/atrophy refers to the death of cells without replacement (e.g., Parkinson's disease).

2. Division-obsessed cells lead to various cancers.

3. Death-resistant cells negatively impact our immune systems as we age, causing us to be more susceptible to illnesses.

4. Mitochondrial mutations may eventually cause disease. Mitochondrial DNA is only inherited from the mother, and each mitochondrial organelle typically contains multiple mitochondrial DNA (mtDNA) copies. During cell division, the mtDNA copies segregate randomly between the two new

mitochondria, and then those new mitochondria make more copies. If only a few of the mtDNA copies inherited from the mother are defective, mitochondrial division may cause most of the defective copies to end up in just one of the new mitochondria. Mitochondrial disease may become clinically apparent once the number of affected mitochondria reaches the threshold of expression.

5. Intracellular junk is molecular garbage in the cells that leads to diseases like cardiovascular disease and macular degeneration. In the case of cardiovascular disease, white blood cells that process cholesterol are unable to handle oxidized cholesterol, causing a buildup within the arteries.

6. Extracellular junk includes accumulations of sticky, malformed proteins that no longer serve their function but instead impair cell or tissue function by their presence.

7. Extracellular crosslinks—that is, changes in the extracellular matrix—are a form of damage that occurs among structural proteins and impairs their ability to move. One example is stiffening of our arteries, which leads to cardiovascular disease.

The idea behind this approach is to treat old age as we would any other medical dilemma, as aging is simply a side effect of living longer. Therefore, the goal is to maintain the damage before it becomes major.

Physical Changes

We have examined our core makeup—our cells and DNA. Now we will expand our understanding of how these changes on the molecular level impact our organs and other physical structures. We have grasped that as the number of healthy cells decreases, our organs' ability to function tends to decline as well. The majority of our bodily functions typically begin to deteriorate just after we reach the age of thirty; however, most functions remain adequate for our chronological age. The decline in function may simply mean that certain stressors, such as physical activity, extreme changes in one's environment, and disorders, take a more significant toll than before. We have broken down the changes to help you grasp the reasons behind the physical manifestations that we see as we age. (This section is adapted from *The Merck Manual: Home Edition*, "The Aging Body," available at http://www.msdmanuals.com/home/older-people-s-health-issues-the-aging-body/changes-in-the-body-with-aging.)

Mental Function

One of the most common traits of older people is their difficulty remembering or coming up with the correct word; we see this during conversations with our grandparents as they recall a story from a few weeks before. Their ability to recall either recent or past events can cause them great frustration. Sometimes they may also find it challenging to learn new things or even to concentrate on the task at hand. The nerve cells in the brain release different amounts of chemical messengers, which send impulses from cell to cell, thus allowing us to recall information, focus, and store new information. As we age, the number of receptors on the nerve cells may decrease, causing our brains to process impulses poorly or more slowly. In some instances, the messages between receptors may not be sent at all.

Physical Activity

Changes in physical activity involve many factors as areas such as the muscles, heart, and lungs decreasing in their functionality. Let's first begin with the muscles. Aging and loss of muscle strength go hand in hand along with difficulty moving and poor flexibility. Muscle quality deteriorates due to a decrease in the number and size of muscle fibers; a reduction in growth hormone; and—for men—less testosterone. As a result, our bodies lose muscle tissue, and it is replaced by fatty or fibrous tissue, which manifests as decreased strength and stiffer muscles. Flexibility and movement are related to a decrease in the amount of joint fluid that the body produces. The cartilage between the bones in the joints also becomes stiffer and may erode, while our tendons and ligaments also become stiffer and weaker. The way we walk and stand may change due this newly developed erosion. The loss of valuable calcium and other minerals from our bones' infrastructure also is a contributing factor. The loss of these vital nutrients accounts for many of the changes we mentioned above, which can lead to severe damage as we age; hip replacements and knee surgeries are commonplace among the mature crowd. Loss of balance or unsteadiness also occurs with age when the structures in the inner ear that help with balance stiffen and deteriorate slightly. Also, the part of the brain that controls balance, the cerebellum, may degenerate over time.

Our hearts are also being modified structurally with the onset maturity. The valves that control the direction of blood flow become thicker and stiffer, sometimes due to an excessive buildup of fat and cholesterol within the arteries, which can cause them to narrow or become clogged. This has a domino effect on the vessels, preventing the heart from pumping adequate blood throughout the body. The stiffness of the blood vessels prevents them from constricting to force the blood out from the lungs. As a result, when we stand, there is insufficient blood pressure to maintain our stance. Our nervous system, which signals the heart to increase blood flow, is also not functioning as effectively, which doubly impacts the flow of blood to our head, leading to dizziness and light-headedness. Strenuous exercise becomes a challenge. We have already mentioned the heart's inability to keep up with demand, which is particularly evident during exercise. The heart cannot speed up as quickly or pump as fast as it used to. In addition, the chemical messenger that would normally stimulate it no longer has the same effect as in youth. Aging also takes a toll on the lungs, as they are unable to keep up with the demand for oxygen during exercise.

Our Senses

Age affects eyesight, hearing, smell, and taste. Vision changes in numerous ways. The lens of the eye stiffens, making it difficult to focus on close objects, leading us to rely on reading glasses. The retina becomes less sensitive to light. The lens of the eye becomes less transparent, the pupils react more slowly to changes in light, and the darkened areas of the lens increase glare. All these factors contribute to poor vision in dim light and the challenges exhibited in changing light levels. Dry eyes manifest due to a decrease in the number of cells that help to lubricate our eyes and a reduction in tears produced by the tear glands. Age-related hearing loss, known as presbycusis, is another cumulative effect of aging and is typically noted at high frequencies. The lining of the nose gets thinner and drier, and the nerve endings in the nose also deteriorate, leading to a poor sense of smell. Lastly, our affinity to taste also diminishes as our taste buds become less sensitive and we produce less saliva.

Changes in our senses and additional physical changes within the mouth may alter our eating habits. Dryness in the mouth and weakened jaw and throat muscles impair coordination and can make swallowing a challenge. The bones at the top of the spine shift their positioning, causing the head to tip forward and compressing the throat. To add to this, because our senses are no longer stimulated and we lack moisture in our mouths, food may become less appetizing for some.

Skin and Hair

Wrinkles, gray hairs, and even loss of hair are all traits of old age. The fat layer under the skin that acts as both a cushion and insulation becomes thinner with age. As a result of this, the body is unable to adjust to changing temperatures as easily. The reduction in the number of blood vessels also plays a role in temperature regulation, since blood flow within the deeper layers of the skin decreases, leaving the body unable to efficiently dissipate heat. The number of sweat glands diminishes, adding to the body's instability to manage higher temperatures, as sweat helps to cool the body. Collagen and elastin in the skin are produced in smaller amounts, leading to reduced suppleness, which both creates wrinkles and can lead to tearing of the skin. Additional bruises and broken blood vessels are evident as blood vessels in the skin become more fragile. Wounds are slower to heal due to the reduction in the number of blood vessels in the skin, and the cells responsible for healing wounds are also reduced and act more slowly.

Gray or white hair is linked with the reduced production of the pigment melanin, which provides the hair follicles with their color. Hair must be replaced periodically. However, some of the hair follicles stop producing new hair; therefore, hair tends to grow more slowly or not at all. The number of nerve endings in the skin decreases, lessening sensitivity.

Sexual Function

Healthy aging brings about typical changes in sexual performance for both men and women. For both sexes, a reduction in the flow of hormones directly impacts sexual satisfaction; men experience a reduction in testosterone and women a decrease in estrogen and progesterone production. How does this impact the arousal cycle? The initial phase of arousal is excitement. Men have a slower response and require more manual stimulation to obtain an erection, and the erection is not as rigid as before. This is due to the decrease in blood flow to the penis; there are other risk factors that can contribute to poor blood flow to the penis, including poor diet, a sedentary lifestyle, and being overweight. Women are also slow to reach arousal, and with reduced elasticity and less lubrication within the vagina, intercourse may be painful. The orgasm phase is shorter and weaker for both sexes, with men experiencing a reduction in semen and a less forcible ejaculation.

Societal and Cultural Age Variations

Aging is not only a biological process but also a cultural one. Age comes with certain expectations and privileges that vary across the globe. For example, in most parts of the Caribbean and most of the United States, the age of adulthood is eighteen years. In some Canadian territories, the age is nineteen; in Scotland, it is sixteen; and in Japan, it is twenty. At the age of adulthood, the coat of childhood play must be shed to be replaced by the maturity of adulthood. It should be noted that in some East Asian countries, such as Korea, age is calculated differently than in the West. In Korea, age does not begin with the birth of a child but starts at the beginning of the year, rounding up the time spent by a child in the mother's womb. People become older not on the day of their birth but on January 1 according to the lunar calendar. For example, a child born in late December 2013 would turn two on January 1, 2014.

Adulthood and old age can have both negative and positive connotations depending on one's cultural background. Perceptions of aging influence societal behaviors and expectations toward older people as well as older adults' well-being and coping with the aging process. Depending on cultural and personal philosophy, aging can be seen as an undesirable phenomenon that reduces beauty and brings one closer to death or as an accumulation of wisdom, a mark of survival, and a status worthy of respect. In many instances, loss of youth is seen as a loss in value, and thus a stigma is created surrounding aging. This is particularly true for women, who tend to be seen as older earlier than men; older men are typically portrayed as more desirable.

Western culture considers the elderly a burden. However, recent statistics show a significant increase in the number of people over age sixty-five, which is set to continue to grow exponentially due to improved technology leading to longer life spans. The focus now is on retaining the productivity of an aging population while minimizing the significant economic impact that it currently has on health care and governments.

What Is Successful Aging?

The idea of successfully aging or "aging gracefully" has been in existence for a few decades. In earlier research, the focus was on the disabilities that come with age and how they could be managed. However, the progress made in gerontology studies has now expanded their research to understand why we age and how we can slow down or stop the process. Successful aging can be defined based on three factors. The first is a low probability of disease or disability, the second is a high cognitive and physical functioning capacity, and the final factor is active engagement with life.

With that being said, there is an array of theories that consider the societal aspects of aging.

1. Disengagement theory states that the separation of older persons in society is normal and appropriate; this separation is beneficial both to society and to the older individuals. Although this theory originally received a lot of attention, it has been greatly criticized for obvious reasons.

2. Activity theory is the idea that the more active elderly persons are, the more likely they are to be satisfied with life.

3. Selectivity theory is a synthesis the two theories above. It suggests that it may be beneficial for older persons to become more active in some aspects of their lives and more disengaged in other aspects.

4. Continuity theory is the notion that later in life, persons make adaptations that enable them to gain a sense of continuity between the past and the present. This continuity helps contribute to their well-being later in life.

Aging is a global phenomenon that is costing billions of dollars in both health care for and maintenance of the elderly. Statistics show that by 2040, the US government will be spending on average 34 percent of its gross domestic product on the elderly, and it is said that we will soon have more elderly people than children in the world due to low fertility rates (The Economic Case for Health Care Reform 2009). The outlook on health is a challenge with the significant rise in chronic noncommunicable diseases, such as heart disease, diabetes, and cancers, which reflect not only an aging population but also changes in lifestyle and diet. The aim is to create an aged population that is in good health and ideally able to contribute to their communities and families. The economic burden of ill health takes a toll not only on the government but also on individuals who either have poor health or are taking care of loved ones who are ill. It is time for us all to begin to write our own prescriptions to improve health beginning in our youth and continuing well into our mature years.

Unseen forces are the greatest mystery to man.

—Alfred Sparman, MD

2

Metabolism

The term *metabolism* is constantly thrown around. You probably have heard people say, "Her metabolism is young; she can afford to eat anything she wants," or, "My metabolism is slowing down now that I am old." But do we truly understand what metabolism means and what it does for our bodies? The word *metabolism* is derived from the Greek word *metabolismos,* which means "change," and it was first documented centuries ago. Our bodies are under constant and considerable amounts of stress as they try to produce energy from two vital elements—that is, glucose from food and oxygen from the air. This biological process is called metabolism, and it encompasses every bodily function imaginable, including breathing, blood circulation, the contraction of our muscles, waste elimination, body temperature control, and operating our brain and nerves. Metabolism refers to all chemical reactions that occur in living organisms, which not only includes digestion but also the transport of substances into and between different cells. The latter is termed intermediary or intermediate metabolism.

Metabolism is broken down into two separate systems that work in synch with each other. The first is catabolism, and the second is anabolism. Catabolism breaks down organic matter, such as our food, and harvests energy via cellular respiration. Anabolism uses the energy produced to construct components for the cells, such as proteins. Metabolic pathways are the avenues that the chemical reactions of metabolism follow. One chemical is transformed through a series of steps into another chemical with the help of a sequence of enzymes. Enzymes are the drivers, as they help to connect events that would not occur on their own; they create the spontaneous reaction that allows for the release of energy. They are the catalysts and instigators for the reactions as well as the regulators of the metabolic pathways in response to changes in the cell's environment or to signals from other cells. Our metabolic system is the gatekeeper that determines which substances are nutritious and which are toxic and thus need to be removed. Additionally, our metabolic rate, the speed of our metabolism, influences the amount of food we need to survive.

To understand how our metabolic structure works, we will need to learn about the key biochemicals and minerals that are vital for life.

Biochemicals and Minerals

Our bodies are made up of structures that are based on three basic classes of molecules: amino acids, carbohydrates, and lipids (what we typically refer to as fats). Metabolic reactions focus on creating these three molecules or breaking them down to be used as a source of energy via digestion. These biochemicals can be joined together to make polymers that are the essential macronutrients of life.

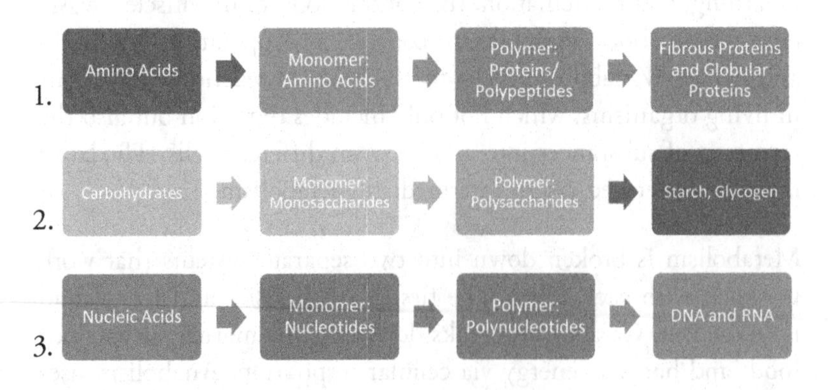

Amino Acids and Proteins

Amino acids create proteins by joining together in a linear chain linked by peptide bonds. Proteins perform multiple functions in the body. As enzymes that catalyze the chemical reactions in metabolism, they have structural or mechanical functions. For example, proteins are the scaffolding that helps cells to maintain their shape. Cell signaling, immune responses, cell adhesion, active transport across cells, and assisting with the cell cycle—particularly when glucose is low—are all key jobs for the protein molecule.

Lipids

The main purpose for this molecule is as part of biological membranes (such as the cell membrane) both internally and externally. Lipids are also a source of energy. Cholesterol is an example of a major class of lipids.

Carbohydrates

These are the most abundant molecules and fill numerous roles, such as the storage and transport of energy and structural components. The basic units or monomers are called monosaccharides and include fructose and glucose. These monomers join together in almost limitless ways to form polysaccharides.

Nucleotides

DNA and RNA are two nucleic acids that are polymers of nucleotides. Each nucleotide is composed of a phosphate group with a nitrogenous base. Storage and use of genetic information and interpretation via transcription and protein synthesis are the crucial roles for nucleic acids. Nucleotides also act as coenzymes in the metabolic-group-transfer reaction.

Coenzymes

Coenzymes are the group-transfer intermediaries that allow cells to use a small set of metabolic intermediaries to carry chemical groups between different reactions. Each class of reactions is carried out by a specific coenzyme, which is a substrate for the set of enzymes that they produce and those enzymes that will consume it. These coenzymes are constantly made, consumed, and recycled. Adenosine triphosphate (ATP) is one of the main coenzymes and the universal currency for the energy of the cell. ATP is used to transfer chemical energy between different reactions. It is the bridge between catabolism and anabolism; the catabolic reactions create the ATP, and the anabolic ones consume it. The quantity of ATP in the cells is small, and therefore it must be continuously regenerated.

Vitamins

A vitamin is an organic compound that cannot be made in the cells but is needed in small quantities, as these vitamins function as coenzymes after modification.

Minerals

Minerals are inorganic elements that are needed for metabolism because approximately 99 percent of our mass is made up of the elements carbon, nitrogen, calcium, sodium, chlorine, potassium, hydrogen, phosphorous, oxygen, and sulfur. The majority of carbon and nitrogen is reserved for the organic compounds, proteins, lipids, and carbohydrates, while the oxygen and hydrogen are present in water.

Cell Respiration

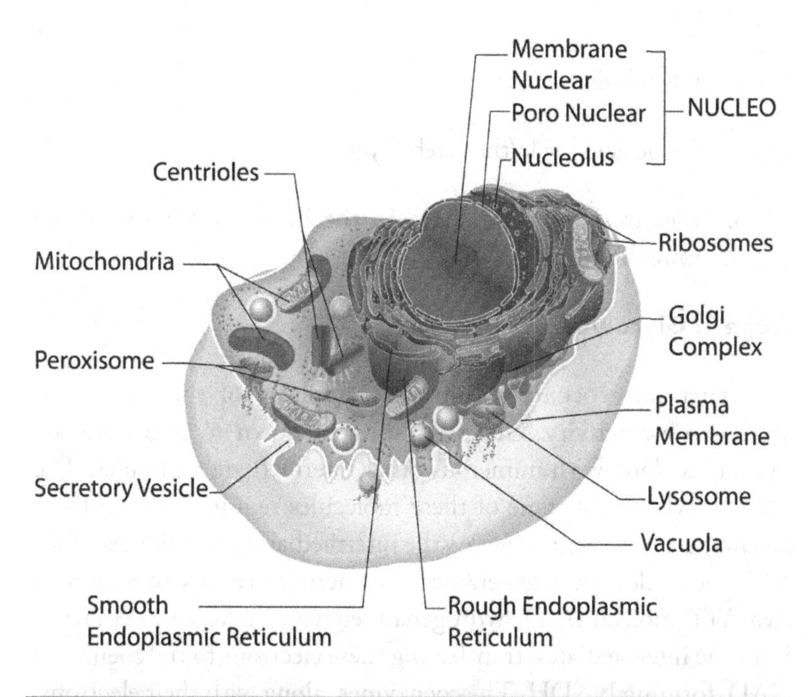

Cells respire to produce the energy necessary for the cell's activities; this energy is harvested from an organic substance—glucose, for example. The metabolic pathways are confined to the membranes; the pathways can either occur on the membranes or in the inner compartments that the membranes may create.

Membranes create a large surface area and small compartments, allowing for high reaction rates to occur. The energy is created when the end phosphate is released, thus creating adenosine diphosphate (ADP); the phosphate released is a low-energy molecule. A low-energy process adds on a phosphate to re-create ATP. ATP can be transferred to other molecules involved in cellular processes, such as DNA replication, active transport, synthetic pathways, and muscle contraction. There are three successive pathways of cellular respiration:

1. Glycolysis
2. Citric acid cycle/the Krebs cycle
3. Electron transport chain in the inner membrane of the mitochondria

Stage 1: Glycolysis

This first stage occurs within the cytoplasm of the cell and is an anaerobic activity. Glucose is broken down to two pyruvate molecules along with numerous other intermediate molecules. The phosphorylation of some of these molecules requires two ATPs in an energy-investment stage. As the intermediates are produced, four ATP molecules are regenerated. This activity results in a gain of two ATP molecules. Dehydrogenase enzymes take away electrons from the intermediates, transferring these electrons to the coenzyme NAD, forming NADH. The coenzymes, along with their electrons, are transported to the inner membrane of the mitochondria to be used in the electron transport chain.

Once oxygen is available, the pyruvate molecules will continue on to the citric acid cycle. If there is no oxygen, the pyruvate will undergo lactic acid fermentation, producing lactic acid.

Stage 2. The Citric Acid Cycle

The citric acid cycle occurs as the pyruvate enters the matrix of the mitochondria and the carbon dioxide is removed. The removal of these carbon molecules forms an acetyl group that combines with coenzyme A to form acetyl coenzyme A. This then combines with a molecule called oxaloacetate to form citrate—which is why we call this process the citric acid cycle. The citrate is then broken down using a series of enzyme-controlled reactions to regenerate oxaloacetate. As a result of the removal of carbon in the form of carbon dioxide and hydrogen, intermediate molecules form. Carbon dioxide is a by-product of this process.

In addition, dehydrogenase enzymes remove hydrogen ions and electrons from intermediates, which are passed to coenzymes NAD or FAD (forming NADH or $FADH_2$). These high-energy electrons are passed to the electron transport chain.

Stage 3. The Electron Transport Chain

The last stage in this energy-creating respiration pathway is the electron transport chain, which produces the most ATP molecules. It is a collection of proteins that are found in the inner membrane of the mitochondria. NADH and $FADH_2$ release electrons into the transport chain. The electrons transfer their energy to the proteins that are in the membrane, providing the energy for the hydrogen ions to be pumped across the membrane. ATP is synthesized by the flow of ions across the membrane with the assistance of a protein called ATP synthase. Three ATP are produced for each NADH, and two ATP are produced from each $FADH_2$, which then transfers high-energy electrons to the electron transport chain. The total gain from the electron transport chain is thirty-four ATP molecules. The final receptor is oxygen, which combines with hydrogen to form water. If the oxygen is not present, then the hydrogen will not be able to pass through the electron transport chain, reducing the number of ATP molecules produced. One glucose molecule yields thirty-eight ATP molecules.

In the event that glucose is not available for the respiration pathway, other respiratory substrates can be used via alternative metabolic pathways. Glycogen, starch, proteins (amino acids), and fats can be broken down into intermediates in glycolysis or the citric acid cycle. This provides alternative pathways to make ATP.

Catabolism and Anabolism

Catabolism breaks down large molecules, which includes food molecules, into smaller units and oxidizes them. The purpose of catabolism is to create the energy that is needed for anabolic reactions. The most common group of catabolic reactions can be broken into three stages. In the first stage, large organic molecules (proteins, polysaccharides, or lipids) are digested into smaller components outside of their cells. The cells take up these smaller molecules and convert them to even smaller molecules that release some energy, usually acetyl coenzyme A (acetyl-CoA). The final stage involves oxidizing the acetyl group on the CoA to water and carbon dioxide in the citric acid cycle and the electron transport chain, releasing the energy that is stored by reducing the coenzyme nicotinamide adenine dinucleotide (NAD+) into NADH.

Digestion

Digestion is one of the catabolic reactions where macromolecules, such as starch or proteins, cannot be rapidly taken up by the cells and therefore must be broken down into their smaller units in order to be used in cell metabolism. Different classes of enzymes digest these polymers. For example, proteases convert proteins to amino acids, and glycoside hydrolases digest polysaccharides into their simple sugars, monosaccharides. Active transport proteins pump into cells the amino acids or sugars that are released in digestion.

Energy from Organic Compounds

Organic compounds include carbohydrates, fats, and amino acids; these compounds are all sources of energy once catabolized. Carbohydrates are catabolized and broken down into monosaccharides, which are usually taken into the cells. Once inside of the cells, the monosaccharides are further broken down, usually via glycolysis. These simple sugars, such as glucose and fructose, are converted to pyruvate and also produce a small amount of ATP. Some of the pyruvate functions as an intermediary in many other metabolic pathways; however, the majority is converted to acetyl-CoA for consumption in the citric acid cycle. The products of this cycle are ATP and, most importantly, NADH, which is created from the oxidation reaction of acetyl-CoA from NAD+. Carbon dioxide is released as a waste by-product of this oxidation process. In conditions where oxygen is absent, glycolysis will produce lactate via the enzyme lactate dehydrogenase, which causes the NADH to be reoxidized to NAD+ for recycling in the glycolysis cycle. Glucose can also be broken down via the pentose phosphate pathway. Via this route, the coenzyme NADPH is reduced and produces pentose sugars (e.g., ribose) that are the sugar component of nucleic acids.

Fats are broken down into free fatty acids and glycerol via hydrolysis; the glycerol heads for the glycolysis cycle, while the fatty acids are disassembled via beta oxidation to release acetyl-CoA, which is then pumped into the citric acid cycle. The energy released from fatty acids after oxidation is greater than that released from carbohydrates, because carbohydrates contain more oxygen in their structure.

Amino acids are used either to create proteins and other biomolecules or are oxidized to urea and carbon dioxide as an alternative source of energy. The first step in the oxidation of amino acids involves a transaminase removing the amino group. This amino group is sent to the urea cycle, leaving behind a deaminated carbon skeleton in the form of a keto acid. Many of these keto acids are intermediaries in the citric acid cycle. An example of this is the deamination of glutamate, which forms alpha-ketoglutarate. Gluconeogenesis allows for these glucogenic amino acids to be converted to glucose.

Energy Transformations: Oxidative Phosphorylation

Oxidative phosphorylation is "the metabolic pathway in which the mitochondria in cells use their structure, enzymes, and energy released by the **oxidation** of nutrients to reform ATP" (Berg et al. 2002). ATP forms as a result of the transfer of electrons from NADH or $FADH_2$ to O_2 by a series of electron carriers. This occurs as part of cell respiration in the mitochondria and is the process by which ATP is created through the synthesis of ADP. However, the end product of oxidative phosphorylation is reactive oxygen species that contribute to the creation of free radicals (Berg et al. 2002).

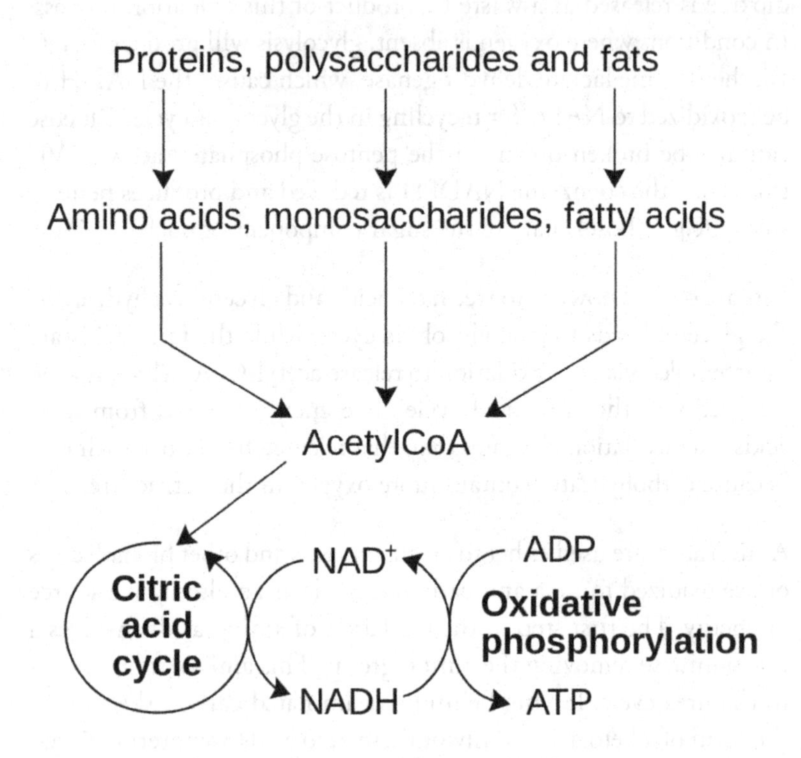

Anabolism

Anabolism is the other half of metabolism where the energy that is released from catabolism is used to synthesize complex molecules. The complex molecules in our body are built step-by-step from small and simple precursors. Anabolism is divided into three elementary stages. There is a preproduction of the anabolic precursors amino acids, monosaccharides, isoprenoids, and nucleotides. These precursors are activated to their reactive forms using ATP for energy. Finally, these precursors are assembled into complex molecules, such as proteins, polysaccharides, lipids, and nucleic acid.

Carbohydrates

Carbohydrate anabolism is where simple organic acids can be converted into monosaccharides, such as glucose, and used to assemble more complex polysaccharides like starch. Gluconeogenesis is the name given to the process that allows for the generation of glucose from pyruvate, lactate, glycerol, glycerate 3-phosphate, amino acids, and other such compounds. In this process, pyruvate is converted to glucose-6-phosphate via a chain of intermediates. The gluconeogenesis pathway is not simply the reverse of glycolysis, as nonglycotic enzymes catalyze many of the steps. This prevents the two pathways from running simultaneously, so the formation and breakdown of glucose are regulated independently.

Fat is the usual way that energy is stored. However, in humans, fatty acid stores cannot be converted to glucose via the gluconeogenesis pathway. Humans are unable to convert acetyl-CoA to pyruvate due to the absence of the necessary enzymatic machinery; plants, on the other hand, are able to. Therefore, after a period of extended starvation, vertebrates need to create ketone bodies from fatty acids to replace the glucose in tissues, such as the brain. Conversely, plants and bacteria use the glyoxylate cycle, which skips the decarboxylation step in the citric acid cycle, allowing the transformation of acetyl-CoA to oxaloacetate. This product can then be used for the production of glucose.

Polysaccharides and glycans are complexes of simple sugars and are created by glycosyl transferase from a reactive sugar-phosphate donor to an acceptor hydroxyl group on the growing polysaccharide. The polysaccharide can have either straight or branched structures, because any of the hydroxyl groups along the ring of the substrate can be acceptors. These complexes either can have structural or metabolic functions or can be transferred to lipids and proteins by an enzyme called oligosaccharyltransferases.

Fatty Acids

Fatty acid synthases produce fatty acids via the polymerization and then reduction of acetyl-CoA units. Fatty acids' acyl chains are lengthened by a cycle of reactions that add the acyl group, reduce it to an alcohol, dehydrate it to an alkene group, and then reduce it once again to an alkane group. There are two types of enzymes that are a part of fatty acid biosynthesis; for animals and fungi, a single, multifunctional, type I protein performs the reaction. For plant plastids and bacteria, there are separate type II enzymes that perform each step in the pathway.

Carotenoids are an example of terpenes and isoprenoids, a large class of lipids that form the largest class of plant natural products. As a result of the assembly and modification of isoprene units from the reactive precursors of isopentenyl pyrophosphate and dimethylallyl pyrophosphate, these compounds are created. In animals and archaea, the precursors are formed from the mevalonate pathway, which produces them from acetyl-CoA. In plants and bacteria, the nonmevalonate pathway instead uses pyruvate and glyceraldehydes 3-phosphate as a substrate for the precursors.

Proteins

Humans can synthesize eleven of the twenty most common amino acids; we must obtain the remaining nine amino acids from food. Amino acids are manufactured from the intermediaries in glycolic acid, the citric acid cycle, or the pentose phosphate pathway. Glutamate and glutamine provide the necessary nitrogen. Peptide bonds are the glue that holds the chain of amino acids together to make proteins. There is a unique sequence of amino acids for each protein; the amino acids can be combined in numerous ways to create an endless amount of proteins. The different shapes and sequences give the proteins different functions. For example, keratin is a fibrous protein found in hair and nails.

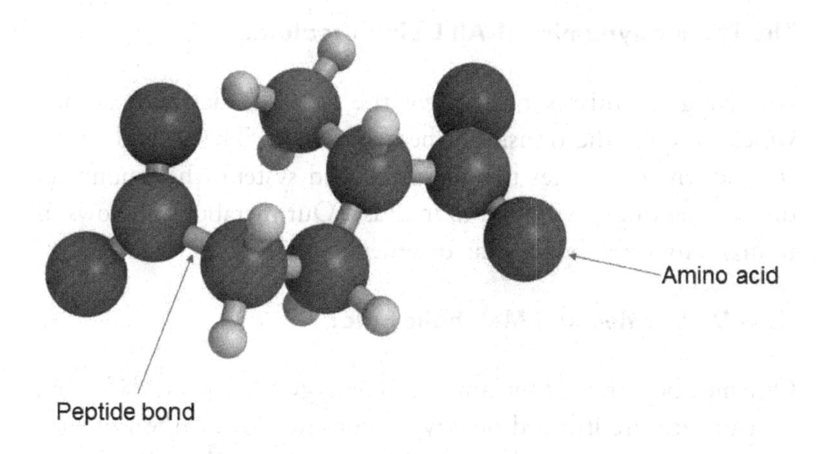

Nucleotide Manufacture

Nucleotides are also made from amino acids, along with carbon dioxide and formic acid, in pathways that require a large amount of metabolic energy. As a result, our bodies have created a salvaging system for these nucleotides to minimize the energy deficit.

Xenobiotics and Redox Metabolism

Our cells face two dilemmas that require special enzymes and metabolites to regulate: the buildup of toxic compounds and oxidative stress. Xenobiotics are the damaging compounds that result from exposure to compounds that cannot be used as food and can be harmful if accumulated in our cells, such as synthetic drugs, natural poisons, and antibiotics. A group of xenobiotic-metabolizing enzymes efficiently detoxifies these nonfunctional compounds. The second challenge is oxidative stress due to the fact that we are aerobic creatures and require oxygen to survive. During the formation of disulfide bonds during protein folding and oxidative phosphorylation (the reformation of ATP), reactive oxygen species, such as hydrogen peroxide, are produced. Antioxidant metabolites and specialized enzymes remove this type of oxidant.

The Thermodynamics of All Living Creatures

All living organisms must obey the laws of thermodynamics, which describe the transfer of heat and work. The second law of thermodynamics states that in any closed system, the amount of disorder (entropy) will tend to increase. Our metabolism allows us to maintain order by creating disorder.

How Do We Measure Metabolic Rate?

Our metabolic rate is the amount of energy we expend daily, and we can measure it based on oxygen consumption, carbon dioxide production, or heat production. The rate calculated while at rest is called the basal metabolic rate (BMR). An equation that includes your body mass, age, weight, and height allows you to determine your BMR. Your gender also affects BMR, as men typically have a higher metabolic rate than women. BMR is measured in kilojoules (kJ). The Harris-Benedict equation is used to calculate your BMR, and the resulting number is the daily kilocalorie intake you need to maintain your current body weight:

- For men, BMR = 66.47 + (13.7 × weight) + (5 × height) – (6.8 × age), where weight is measured in kg, height is measured in cm, and age is measured in years.
- For women, BMR = 655.1 + (9.6 × weight) – (1.8 × height) – 4.7 × age).

The larger you are, the higher your metabolic rate will be and the more energy your body will use at complete rest. This is because larger people need more energy to pump blood around the body to keep its systems functioning effectively. Just as a bigger car uses more fuel, a bigger person uses more energy.

How Do Calories Fit In?

The calorie, also known as the kilocalorie or the big C calorie, is the unit of measure for the amount of energy in food; that is, it denotes the amount of energy per unit of mass (calories per gram or calories per serving). Therefore, your BMR will tell you the amount calories required daily to maintain your current body weight.

How Is This Regulated?

Homeostasis is the set of conditions that must be maintained and regulated by the reactions of metabolism. This regulation allows us to respond to signals and actively interact with our environment. To understand how this works, we need to look at two closely linked concepts: (1) the regulation of an enzyme in a pathway is how its activity is increased and decreased in response to signals; and (2) the control exerted by this enzyme is the effect that these changes in its activity have on the overall rate/flux of the pathway (Soyer, Salathé, and Bonhoeffer 2006). There are different levels of metabolic regulation; intrinsic regulation is where the metabolic pathway self-regulates in response to changes in the levels of substrates in its system. Extrinsic control involves a cell modifying its metabolism in response to signals that usually come in the form of soluble messengers, such as hormones and growth factors from other cells (Fell and Thomas 1995). Specific receptors on the cell's surface receive these messages, and a secondary messenger system, which usually involves the phosphorylation of proteins, transmits them inside of the cell.

A prime example of extrinsic regulation occurs with the regulation of glucose metabolism by insulin. Insulin, a hormone, is produced in response to a rise in blood glucose levels. It then binds to insulin receptors on the cells, activating a cascade of protein kinases that cause the cells to collect the glucose to convert it into storage molecules, such as fatty acids and glycogen.

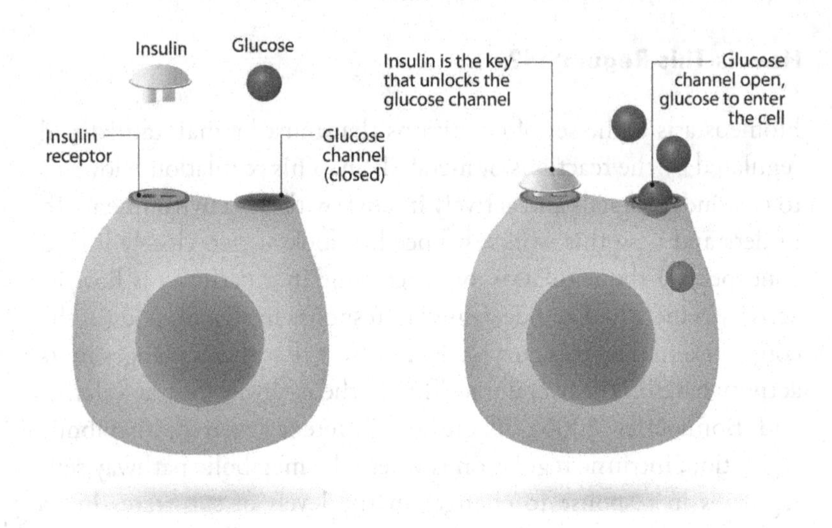

Can This Process Be Inhibited?

Metabolism is a synchronous reaction between molecules and enzymes. However, these enzymes can be influenced by different chemicals, affecting their rates of reaction and possibly gene expression. Inhibitors can stop an enzyme from binding to its substrate. As a result, inhibitors can directly control the progress of a metabolic pathway.

There are three types of inhibition:

1. Competitive inhibition

2. Noncompetitive inhibition

3. Feedback inhibition

Competitive Inhibition

Competitive inhibition occurs when an inhibitor molecule binds to the active site of the enzyme, stopping the substrate from binding. Inhibitor molecules are able to compete with the substrate because they have a similar molecular shape. Increasing the substrate concentration can reverse competitive inhibition. The concentration of the substrate eventually allows the inhibitor to become diluted so that all the enzyme molecules bind to the substrate. For example, sarin, a nerve agent, prevents the proper operation of an enzyme that acts as the body's off switch, for glands and muscles. Without an off switch, the glands and muscles are constantly being stimulated (CDC 2013).

Noncompetitive Inhibition

Noncompetitive inhibition occurs when an inhibitor does not bind to the active site but instead binds to a different part of the enzyme and changes the shape of the active site. As a result, the substrate cannot bind to the enzyme, decreasing the reaction rate. Examples of noncompetitive inhibitors include cyanide and mercury.

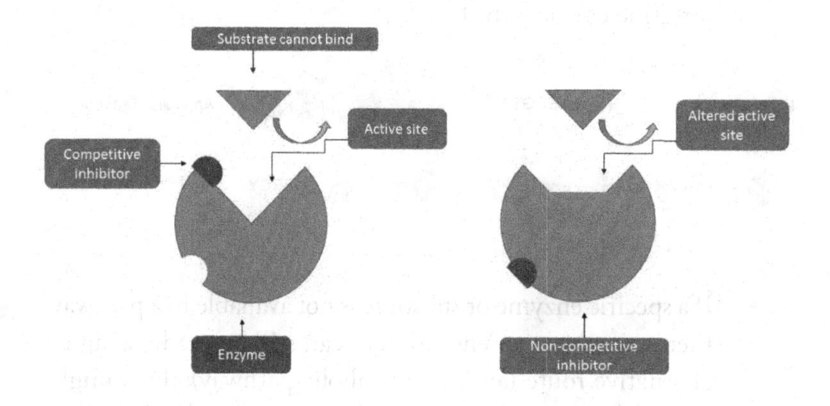

Let us do a quick recap:

- ATP is the energy-carrying molecule used in cells, because it can release energy very quickly.
- Energy is released from ATP when the end phosphate is removed. Once ATP has released energy, it becomes ADP, which is a low-energy molecule.
- Almost all cellular processes need ATP to give reactions their required energy.
- Metabolic currencies remain the same across cells. The metabolic requirements of each cell type are determined by their tissue function and environment.
- Thousands of metabolic pathways exist, and most involve multiple steps—some can be hundreds of steps long!
- Enzymes control metabolic pathways. The enzymes change the substrate at each step in the metabolic pathway in order to get the final product at the end.
- There are different types of metabolic pathways—some are anabolic and some are catabolic. An example of a catabolic reaction is the process of food digestion, where different enzymes break down food particles so that the small intestine can absorb them.

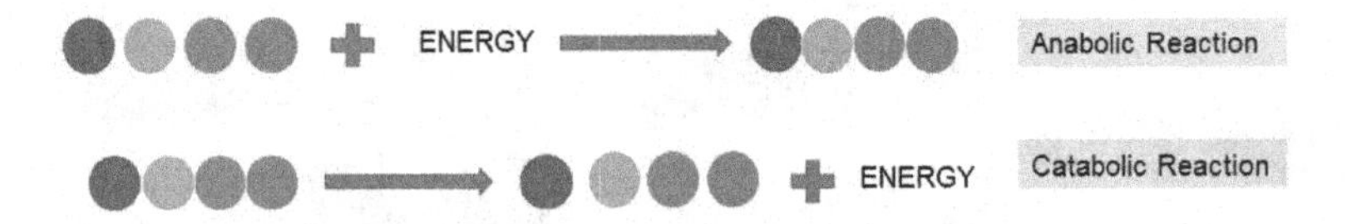

- If a specific enzyme or substrate is not available in a pathway, then sometimes an end product can still be made using an alternative route (another metabolic pathway). This might take longer but still results in the necessary end product.
- Enzymes are able to bind to their substrate because they have an active site.

- Competitive inhibition occurs when two entities with similar molecular structure compete for the active site.
- Noncompetitive inhibition occurs when an area remote from the active site is occupied, altering the active site of the substrate.

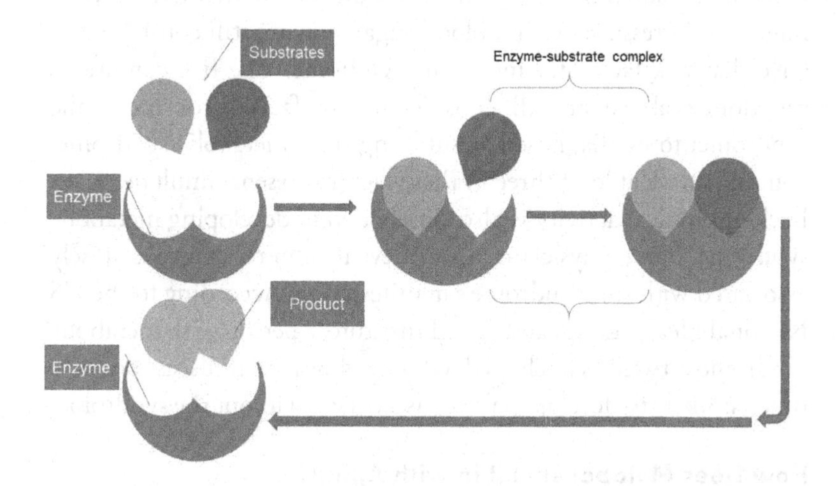

- Although enzymes are specific, some act in groups as multienzyme complexes. This means that some enzymes can act on more than one substrate.

Metabolic Syndrome

Metabolic syndrome is a group of conditions that, when combined, place you at a higher risk for cardiovascular disease, diabetes, and stroke. The conditions include

- abdominal obesity or what would be termed as having a large waistline;
- a high triglyceride level (150 or more milligrams per deciliter of blood);
- a low level (below 40 mg/dL of blood for men and below 50 mg/dL for women) of high-density lipoproteins, which help remove cholesterol from your blood;

- high blood pressure, meaning that the systolic pressure (the top figure) is 130 mm Hg or greater and the diastolic pressure (the bottom figure) is 85 mm Hg or greater; and
- high fasting blood sugar—that is, 100 mg/dL or greater.

It should be noted that if you are on medication to maintain or treat high blood pressure or high blood sugar, you are still considered to have that risk factor for metabolic syndrome. The five conditions mentioned above are all considered risk factors for metabolic syndrome; to be diagnosed as suffering from metabolic syndrome, you must have at least three of these factors present simultaneously. Lack of physical activity is also closely tied to developing metabolic syndrome. Some physicians also believe insulin resistance is closely associated with this syndrome's manifestation. According to the US National Heart, Lung, and Blood Institute, a person with metabolic syndrome is twice as likely to develop cardiovascular disease and five times as likely to develop diabetes as a person without the syndrome.

How Does Metabolism Fit in with Aging?

Metabolic activities are ongoing in our bodies. This can create metabolic stress, which over time can cause damage to our bodies. As noted above, one of the side effects of oxidative phosphorylation is the production of free radicals. The production of these free radicals has been hypothesized to be the leading cause of aging and its side effects. Free radicals also are associated with the development of cardiovascular diseases and cancer. As we proceed, we will delve deeper to understand from the micro level the age-related changes that can lead to the manifestation of illnesses.

Protons and electrons separate, and so do we.

—Alfred Sparman, MD

3

Atoms and Molecules

The atom is most commonly noted for being the smallest indivisible component of matter. These basic building blocks constitute the corporeal reality that we experience environmentally and articulate physically. Their existence is literally everything, and their discovery was a revolutionary historical turning point in the world of science. Physics, chemistry, biology—all these disciplines share a common link, a fundamental rooting in atomic theory. When considering the function, structure, and performance of the human body, it is important to remember that we are more than what we can see with just the naked eye. Electrical energy fuels our movements and processes; we are ongoing chemical reactions! Understanding life at this level gives you access to knowledge that can prolong your life span and increase the quality of that experience.

We begin our atomic journey in ancient Greece with Leucippe of Milet and his disciple Democritus. Long before the scientific method was introduced to prove theories with experimental evidence, philosophers and theologians observed our natural world and deduced hypotheses (educated ideas and questions). "The history of ancient atomism is not only the history of a theory about the nature of matter, but also the history of the idea that there are indivisible parts in any kind of magnitude—geometrical extension, time, etc." (Berryman 2005).

The physical senses provide the assumption that all matter is continuous. For example, we experience the air that surrounds us as a fluid medium; we do not notice its individual particles. The same can be said for water. Considering the assumptions our senses make, it is no surprise that debate would unfold around the existence of atoms. Leucippe and Democritus, known as the fathers of atomic theory, built their model with five pillars:

1. Matter is composed of atoms separated by empty space, through which the atoms move.
2. Atoms are solid, homogeneous, indivisible, and unchangeable.
3. All apparent changes in matter result from changes in the groupings of atoms.
4. There are different kinds of atoms that differ in size and shape.
5. The properties of matter reflect the properties of the atoms the matter contains. ("Leucippus and Democritus" 2004)

Around the year of 460 BC, Democritus coined the term *atom* from the Greek word *atomos* meaning "uncuttable." Although the Greeks receive most of the credit for the theoretical discovery of atoms, during that same period, there existed an Indian philosophy, Vaiseshika, that held the belief that all matter (earth, air, water, and fire) consisted of a finite number of *paramanus,* or atoms. This theory was similar to Aristotle's idea that all matter was composed of the four elements; however, their concept, unlike Aristotle's, included atoms. The Vaiseshika also understood the indestructible nature of *paramanus.* According to them, "Their assemblies into visible things is degradable and, at the end of a worldly period, the atomic bonds dissolve, then after a period of rest, reunite themselves into a new world" (The Birth of the Atom 1999). These philosophers understood the resilient, transformational, and ultimately indestructible nature of atoms. However, due to social bias, the atomic theory was not widely accepted for another two thousand years!

In the early 1800s, people turned their curiosity to questions of matter and its structure. An English meteorologist, physicist, and chemist by the name of John Dalton conducted multiple experiments using gases to formulate his own atomic theory. He based his findings on the concept of atoms and two scientific laws developed late in the eighteenth century to describe chemical reactions:

1. The law of conservation of mass, formulated by Antoine Lavoisier in 1789, states that the total mass in a chemical reaction remains constant—in other words, reactants have the same mass as the products.

2. The law of definite proportions was first proven by the French chemist Joseph Louis Proust in 1799. This law states that if a compound (a combination of two or more different elements) is broken down into its constituent elements, then the masses of the constituents will always have the same proportions, regardless of the quantity or source of the original substance. (Williams 2014)

Dalton expanded on these principles and developed his law of multiple proportions. This law states that if two elements combine at the atomic level to form more than one compound, they achieve this new compound in fixed ratios of small whole numbers. The ratios naturally differ based on the compounds combined, because each atom has its own specific weight. Let's look at tin and tin dioxide, for example. One hundred grams of tin will either combine with 13.5 grams of oxygen or with 27 grams of oxygen. Dalton noticed this pattern repeatedly, and from these findings, he postulated his atomic theory, the first to be based on evidence.

"The first states that elements, in their purest state, consist of particles called atoms. Second, the atoms of a specific element are all the same, down to the very last atom. Third, atoms of different elements can be told apart by their atomic weights. Fourth, atoms of elements unite to form chemical compounds. And finally, atoms can neither be created nor destroyed in chemical reactions, only the grouping ever changes" (Williams 2014). The truly brilliant and remarkable thing to mention about Dalton's model is that all his experiments were done without microscopic equipment! He never physically saw an atom and consequently had no idea of the complex structures that existed within them.

In the late 1800s, J. J. Thomson discovered the electron, a negatively charged subatomic particle. Using a magnet and a cathode ray tube (CRT), a vacuum tube with a fluorescent screen used to view images—think of the bulky part on the back of old television sets—Thomson observed a green ray being produced. The ray was composed of an odd, negatively charged material. After conducting a series of experiments, he discovered that a single one of these negative particles had an atomic mass that was significantly smaller than that of a hydrogen atom—two thousand times lighter! This was a radical discovery, because hydrogen is the lightest element on the periodic table; Thomson concluded that the model previously developed by Dalton had to be incorrect in stating that the atom was an indivisible, discrete unit. He hypothesized that these negatively charged particles, electrons, must come from within the atom and that there must something within the atom with a counter positive charge to create balance. From this came the famous plum pudding model, in which the negatively charged electrons were represented by plums and the surrounding positively charged material was the pudding.

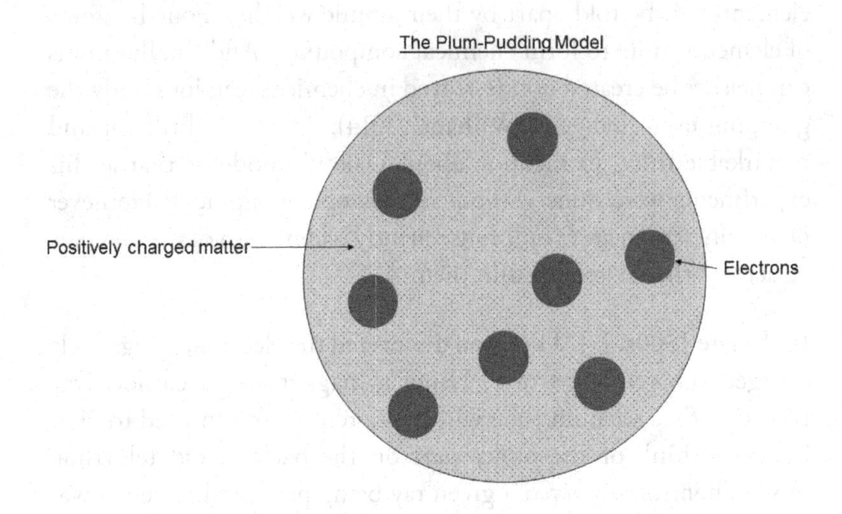

A few years later, a student of Thomson named Charles Rutherford, along with his team—Ernest Marsden and Hans Geiger—ran multiple experiments using alpha particles (positively charged, relatively heavy high-energy particles) and extremely thin sheets of gold foil. Following Dalton's model, the particles should have been able to pass straight through the foil uninhibited. Instead, however, "To the experimenters' amazement, although most of the alpha particles passed unaffected through the gold foil as expected, a small number of particles were deflected at an angle, and a few ricocheted straight back. Rutherford concluded that the atom consisted of a small, dense, positively charged *nucleus* in the center of the atom with negatively charged electrons surrounding it" (Development of the Atomic Theory 2007).

This discovery disproved Dalton's model, which represented atoms as electrons submerged in a dense glob of positively charged matter, creating a unified or symmetrical distribution of energy. Rutherford's model shows us that the atom is mostly made up of space! The electrons move around the nucleus, creating circular orbits.

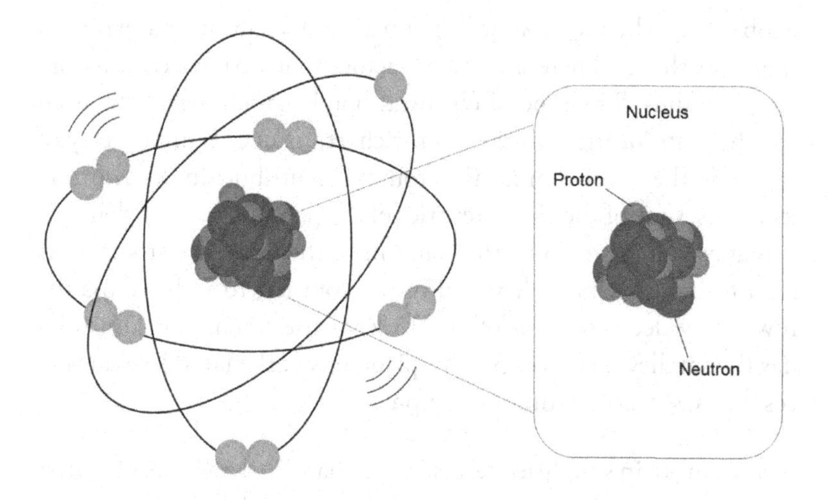

Atomic Structure

Up until this point, the atom was the smallest conceivable division of matter. Subatomic particles are all particles that are smaller than atoms and exist within them. The fundamental constituents are protons and neutrons (they make the dense nucleus) and the miniscule orbiting electrons. Quarks in their six flavors (up, down, charm, strange, top, and bottom) are elementary subatomic particles that make up protons and neutrons; they are but one class of elementary subatomic particles. Others include muons and neutrinos. Quarks are any member of subatomic particles that are affected by the strong nuclear force, a fundamental force of nature (the other fundamental forces are gravitational force, the weak nuclear force, and electromagnetic force). Quarks have no obvious structure, and as far as we know, they cannot be broken down any further. They exist in pairs and can always be found with their antimatter counterparts, antiquarks.

Atoms in general lack a rigid, defined outer boundary or shape. To account for this, atoms' dimensions are usually described in terms of the distance from the nucleus out to their cloud of electrons; this measurement is known as the atomic radius. This principle is observed when atoms are contained in a vacuum and exhibit a spherical shape. There are many factors that influence an atom's atomic radius. The type of chemical bond that connects the atom to others, its location on the atomic chart, its quantum property of spin, and the number of nearby atoms all contribute in determining its radius. One of the characteristics of the periodic table of elements is that as you move down the columns of the table, the atoms' sizes tend to increase, but when you move from left to right across the rows, they decrease. Out of all 118 known elements, helium (He) has the smallest radius of 32 pm (picometers); the largest belongs to cesium (Cs) with a radius of 225 pm.

Atomic mass, in simplified terms, is the quantified measure of matter within an atom of a given element. The term in itself is historically derived from the fact that chemistry was the first discipline to account for the physical properties of things on both a macro and a micro scale. The concept becomes complicated when considering that most substances on the macro level are not pure in form; they contain mixtures of isotopes (chemical variations of a single element), whereas the mass on the micro level is attached to the most common isotope. Because of this, the mass of a physical object will never be equal at both levels. To account for this discrepancy, the mass at the macro level is identified as an atom's molecular weight. "The observed atomic mass is slightly less than the sum of the masses of the protons, neutrons, and electrons that make up the atom. The difference, called the mass defect, is accounted for during the combination of these particles by conversion into binding energy, according to an equation in which the energy (E) released equals the product of the mass (m) consumed and the square of the velocity of light in vacuum (c); thus, $E = mc^2$" ("Atomic Mass" 2014).

Protons

Protons have a positive charge, which means they are made up of two up quarks (+⅔ charge) and one down quark (–⅓ charge), resulting in a positive charge of one. Protons have a mass that is

1,837 times larger than that of an electron, the smallest of the three particles, at 1.6726×10^{-27} kg. "When we say that a proton is made up of two up quarks and a down, we mean that its net appearance or net set of quantum numbers match that picture. The nature of quark confinement suggests that the quarks are surrounded by a cloud of gluons (exchange particles), and within the tiny volume of the proton other quark-antiquark pairs can be produced and then annihilated without changing the net external appearance of the proton" ("Proton").

The amount of protons contained in the nucleus determines the atomic number of an atom. This number is what identifies each individual element. For example, oxygen has an atomic number of eight, meaning it contains eight protons in the nucleus. Inside the nucleus, protons seem to float around the edges, moving extremely fast—faster than any of us can but still much slower than the speed of light. They're held together by the strong force or strong interaction. Particles called gluons are exchanged between quarks and carry this force like a messenger or mediator. The strong force in itself is a fundamental force of nature that underlies and binds all atomic interaction. In chemistry, a proton can simply be referred to or thought of as a hydrogen ion; since it has no neutrons or corresponding electrons, its nucleus is bare, containing only a single proton (Sutton 2014). These $H+$ ions play a pivotal role in how energy is transported within our cells, but we'll discuss that in more detail in the next chapter!

Neutrons

Neutrons are the largest of the three particles, with a mass 1,836 times that of an electron at 1.62929×10^{-27} kg. They have no net charge, containing one up quark and two down quarks. Outside of their difference in charge, protons and neutrons are very similar and often referred to collectively as nucleons.

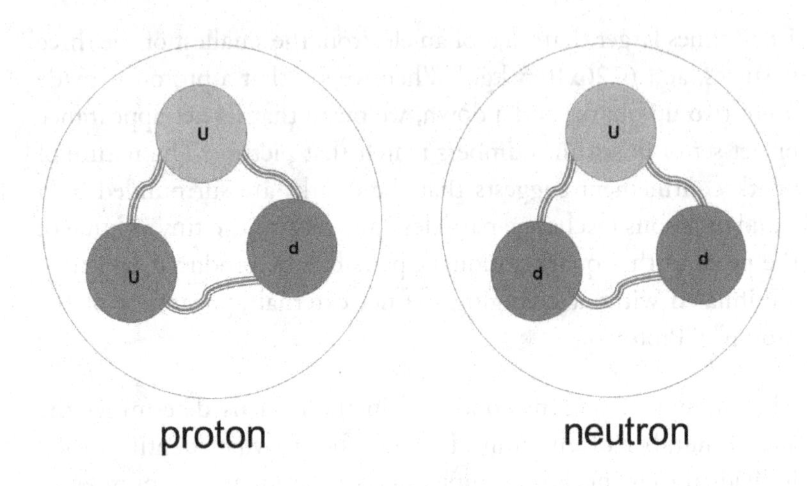

All atoms of the same element will have an identical number of protons; however, the number of neutrons may vary. Elements containing a variation in their neutronic number are called isotopes. Evidence from two separate experiments brought this discovery to light. By 1910, Henry Becquerel had already cemented that certain processes of radioactivity could transform elements. It had earlier been observed that the radioactive elements uranium and thorium, in particular, had small quantities of unidentified substances in their ore that were thought to be new elements and subsequently were given their own individual names, ionium and mesothorium. "Painstaking work completed soon afterward revealed, however, that Ionium, once mixed with ordinary thorium, could no longer be retrieved by chemical means alone. Similarly, Mesothorium was shown to be chemically indistinguishable from radium. As chemists used the criterion of chemical indistinguishability as part of the definition of an element, they were forced to conclude that Ionium and Mesothorium were not new elements after all, but rather new forms of old ones" (Herzog 2014).

In 1910, English chemist Frederick Soddy theorized that elements containing different atomic masses could still be chemically identical and should not be classified as separate substances. Soddy even took it a step further and expanded his findings to both radioactive and stable elements.

Isotopes do not exist at random: "There are 'preferred' combinations of neutrons and protons, at which the forces holding nuclei together seem to balance best. Light elements tend to have about as many neutrons as protons; heavy elements apparently need more neutrons than protons in order to stick together. Atoms with a few too many neutrons, or not quite enough, can sometimes exist for a while, but they're unstable" ("Isotopes"). The stability of the isotope is dependent on the ratio of protons and neutrons. A stable atom is one that has enough binding energy (the total net energy balanced between the strong force and the force of repulsion generated by the interaction of like charges) to hold together the nucleus indefinitely. "For 80 of the chemical elements, at least one stable isotope exists. As a rule there is only a handful of stable isotopes for each of these elements, the average being 3.2 stable isotopes per element. Twenty-six elements have only a single stable isotope, while the largest number of stable isotopes observed for any element is ten, for the element Tin" (Trefil 2015).

An unstable atomic binding energy does not possess this capability and will lose neutrons and protons to achieve stability. This release of atomic energy is radioactive and causes a change within the nucleus. The atom may transmute into a new element to gain stability. The desire to maintain stability is a trait shared by atoms within the body as well. Although we do not run the risk of becoming radioactive, unstable atoms, or free radicals, are highly reactive and can cause chain reactions of damage to cells and the tissues and organs they compose.

Antimatter

The composite materials that exist in particle form and mirror the properties of subatomic particles but have opposing signs are called antimatter. The theory of antimatter first germinated when scientists analyzed and conceptualized the duality and balance between positive and negative electrical charges. "The work of P.A.M. Dirac on the energy states of the electron implied the existence of a particle identical in every respect but one—that is, with positive instead of negative charge. Such a particle, called the positron, is not to be found in ordinary stable matter. However, it was discovered in 1932 among particles produced in the interactions of cosmic rays in matter and thus provided experimental confirmation of Dirac's theory" (Sutton 2015).

How does this work? The antimatter that corresponds with the positively charged proton is the negatively charged antiproton. The antineutron's charge remains negative; however, its magnetic moment (that is, its atomic electromagnetic field) has the opposing sign of the neutron, and the positron is the positively charged counterpart of the electron. The strong attraction between matter and antimatter makes it relatively impossible for them to coexist in close range of each other. In a fraction of a second, the particles will collide with a force so strong they will annihilate each other, a phenomenon that causes a simultaneous disappearance and the release of energy in the form of gamma rays. But it doesn't stop there; there are many other particles and antiparticles.

"By the time the antiproton was discovered, a host of new subatomic particles had also been discovered; all these particles are now known to have corresponding antiparticles. Thus, there are positive and negative muons, positive and negative pi-mesons, and the K-meson and the anti-K-meson, plus a long list of baryons and antibaryons. Most of these newly discovered particles have too short a lifetime to be able to combine with electrons. The exception is the positive muon, which, together with an electron, has been observed to form a muonium atom" (Sutton "Antimatter").

Antimatter is created in the collision of high-energy radiation called cosmic rays. There is currently no evidence that supports the existence of antimatter in large quantities. Our galaxy is dominated by matter, and there are no recorded sites where antimatter and matter collide, producing the gamma rays. The lack of naturally occurring antimatter in the universe goes against Dirac's experimentally supported theory, which states that opposing particles exist equally. If this were the case, matter and antimatter would annihilate each other, and there would be nothingness. This leads to the theory that most particles that existed in the early universe did annihilate each other, and the relatively small amount of particles that survived lacked matching antiparticles and were able to form observable matter. This imbalance is known as matter-antimatter asymmetry.

The Nucleus

All bound protons and neutrons, collectively called nucleons, make up the nucleus of an atom. The nucleus lies at the center of the atom and makes up most of its mass. Nucleons are held together by the attraction of residual strong force. Under these close-range circumstances, this force is much stronger than the repulsive electrostatic force caused by the positively charged protons. According to the Pauli Exclusion Principle, fermions (protons, neutrons, and electrons) cannot occupy the same quantum state at the same time. What this means is that every proton must be in a different state from all other protons. The same goes for all neutrons and electrons. However, a proton and a neutron may occupy the same state. The stability of the nucleus is dependent on its number of neutrons. The actual number of nucleons in the atomic nucleus can be altered through the use of very high energies. In biology, the purpose of the nucleus is to operate as the brain or the command center for the cell. Instructions that dictate the function of the cell are inscribed within the DNA of the cell; the nucleus and the DNA are both protected by a barrier known as the nuclear envelope, which separates these important components from the rest of the cell.

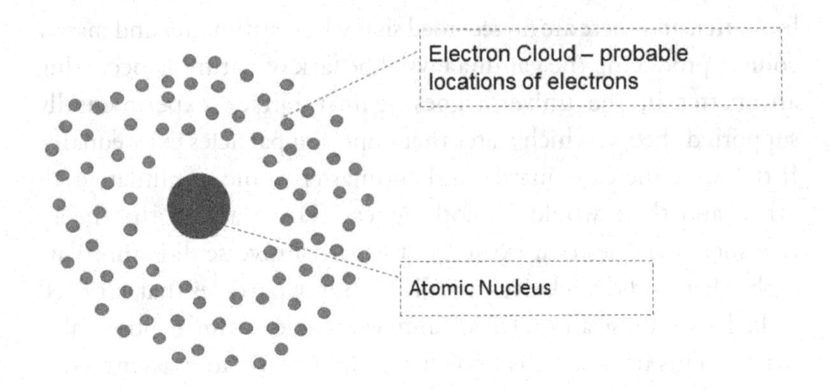

Electrons

The lightest subatomic particle is the electron at 9.11×10^{-31} kg. It is almost massless in comparison to protons and neutrons and behaves like both a wave and a particle. In 1924, a French physicist named Louis de Broglie stated that all matter maintains a de Broglie wave that causes it to behave a lot like light. What this means is that dependent on the conditions, matter—namely electrons—exhibits characteristics of either particles or waves. For example, we can observe the wavelike nature of electrons when we pass a beam of light through parallel slits, which creates something called interference patterns; what you see is an alternating pattern of light and dark bands. "In 1927, Clinton Davisson and Lester Germer observed the diffraction of electron beams from a nickel crystal—demonstrating the wave-like properties of particles for the first time—and George (G P) Thompson did the same with thin films of celluloid and other materials shortly afterwards" (Venugopalan 2010).

An unbound electron has both momentum and mass. It has the ability to consume energy in any amount, but this is not the case for a bound electron. When an electron binds to an atom, it loses some of its characteristics. It can only absorb and emit energy in certain discrete quantities. These energy quantities are referred to as quanta and correlate with photons (the wavelike particles that make up light). Photons within the visible electromagnetic spectrum are associated with color. Consequently, bound electrons play a major role in the colors we experience in nature. "Bound electrons also exhibit various behaviors in magnetic fields. Such interactions are important in analytical chemistry and in the field of medicine" (Summers 2014).

Electrons contain a negative charge. Unlike the previously mentioned atomic particles, they are not made of quarks. Electrons are observably indivisible fundamental particles that are part of a group called leptons. They behave differently from quarks in that they are not affected by the strong force, the atomic force that binds atoms together. They are, however, attracted to the positively charged nucleus due to electromagnetic force. This force tightly binds electrons inside an electrostatic potential well surrounding the smaller nucleus. An outside source of energy is therefore needed for the electron to escape. The strength of the attractive force is completely dependent on how close an electron is to the nucleus. Consequently, electrons that are bound near the center of the electrostatic potential well need more energy to escape than those at greater distances. The electron cloud is the area inside of the potential well where all the electrons form a three-dimensional standing wave, a waveform that does not move in relation to the nucleus. We cannot know for certain where one electron will be at any given moment according to the current model, but atomic orbitals give us the probability of where an electron will appear.

Orbitals are energy levels. An electron can change state to a higher energy level by absorbing a photon with sufficient energy to boost it into a new quantum state. Likewise, through spontaneous emission, an electron in a higher energy state can drop to a lower energy state while radiating excess energy as a photon (Trefil 2015). The amount of energy required to obtain or release an electron is its binding energy. An electron that's bound can only reside in states that are nucleus-centered, and each state is paired with a specific energy level. The lowest energy state of a bound electron is called the ground state (i.e., stationary state), while an electron's transition to a higher level results in an excited state. The electron's energy rises when the quantum number increases because the (average) distance to the nucleus increases. Dependence on the energy of orbital angular momentum is not caused by electrostatic potential of the nucleus but by interaction between electrons (Trefil 2015).

Each orbital has a specific level of energy and specific properties. The orbitals themselves are divided into subshells (related groups), characterized by their angular momentum (the spin of the electron). The subshells are named the S orbital, P orbital, D orbital, and F orbital. Each subshell is paired with a quantum number, which can only be positive numbers. Electrons fill these shells in a way that allows them to minimize the energy being used by the atom. The shells, in increasing order of energy required to fill them, look like this: 1s, 2s, 2p, 3s, 3p, 4s, 3d, 4p, 5s, 4d, 5p, 6s, 4f, 5d, 6p, 7s, 5f, 6d, and 7p. This makes up the electron configuration, the representation of the distribution of electrons within an atom. According to Hund's rule, all orbitals within a subshell have to be filled with a single electron before any shells can hold pairs. The electrons in single occupied shells all spin in the same direction to maximize their net spin. The periodic table is key to understanding the configuration for any element.

The energy level is determined by the period, and the number of electrons is given by the atomic number of the element. Orbitals on different energy levels are similar to each other, but they occupy different areas in space (Hammed 2012). Elements with electrons in the same orbitals are grouped together; for instance, the S blocks are mostly alkali metals, and the D blocks are transition metals.

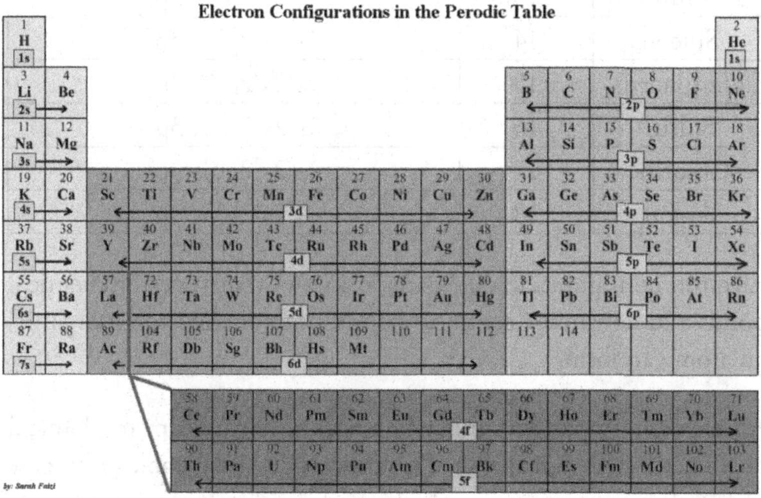

Electron Configurations in the Perodic Table

by: Sarah Faizi

Starting from the nucleus, you count outward until the number of electrons in your shells equals the number of protons (that is, the atomic number) of that element.

Element	Atomic Number	Electron configuration
Hydrogen	1	$1s^1$
Helium	2	$1s^2$
Lithium	3	$1s^2 2s^1$
Beryllium	4	$1s^2 2s^2$
Boron	5	$1s^2 2s^2 2p^1$
Carbon	6	$1s^2 2s^2 2p^2$
Nitrogen	7	$1s^2 2s^2 2p^3$
Oxygen	8	$1s^2 2s^2 2p^4$
Fluorine	9	$1s^2 2s^2 2p^5$
Neon	10	$1s^2 2s^2 2p^6$
Sodium	11	$1s^2 2s^2 2p^6 3s^1$
Magnesium	12	$1s^2 2s^2 2p^6 3s^2$

Aluminum	13	$1s^2 2s^2 2p^6 3s^2 3p^1$
Silicon	14	$1s^2 2s^2 2p^6 3s^2 3p^2$
Phosphorus	15	$1s^2 2s^2 2p^6 3s^2 3p^3$
Sulfur	16	$1s^2 2s^2 2p^6 3s^2 3p^4$
Chlorine	17	$1s^2 2s^2 2p^6 3s^2 3p^5$
Argon	18	$1s^2 2s^2 2p^6 3s^2 3p^6$

Electron configuration allows us to know what types of chemical bonds and how many chemical bonds, both organic and inorganic, an atom can form.

All elementary particles innately possess the quantum mechanical characteristic known as spin. This property is comparable to the case of angular momentum, in which an object is spinning around the center of its mass; the only difference is that with spin, this motion is intrinsic. Technically speaking, the electron does not spin at all; the term is an analogy to the image of a charged spinning object with magnetic properties. Electrons move through a magnetic field, and the way they deflect through this field is similar to that of a charged object—hence the name.

> Unfortunately, the analogy breaks down, and we have come to realize that it is misleading to conjure up an image of the electron as a small spinning object. Instead we have learned simply to accept the observed fact that the electron is deflected by magnetic fields. If one insists on the image of a spinning object, then real paradoxes arise; unlike a tossed softball, for instance, the spin of an electron never changes, and it has only two possible orientations. In addition, the very notion that electrons and protons are solid "objects" that can "rotate" in space is itself difficult to sustain, given what we know about the rules of quantum mechanics. The term "spin," however, still remains. ("What Exactly Is the 'Spin' of Subatomic Particles Such as Electrons and Protons? Does It Have

Any Physical Significance, Analogous to the Spin of a Planet?" 1999)

The spin is measured in units associated with Plank's constant (h). Protons, neutrons, and electrons all have spin +1/2h or spin -1/2h. Electrons that are revolving around the nucleus of an atom possess an additional spin known as orbital angular momentum, while the nucleus possesses its own angular momentum due to its nucleic spin. The magnetic field created by an atom is known as its magnetic moment. This field is produced by the atom's angular momentum and the spin of the paired electrons. Electrons naturally obey the Pauli Exclusion Principle. This principle simply states that no two electrons are to be found occupying the same quantum state at the same time. Each bound electron pair includes one member in a spin up state and another in a spin down state. These spins then cancel each other out.

What this means is that the pairing of electrons is what maintains the neutrality in charge of an atom or molecule. Unpaired electrons are a rare case in chemistry. They happen to be highly reactive with other substances and even with themselves; most are considered unstable unless contained in a vacuum. In organic chemistry, unpaired electrons are known as radicals or free radicals. Free radicals play a very important role in various chemical processes. In biology, some free radicals can regulate cellular processes within the body and are a natural byproduct of metabolic reactions. In the following chapters, we will discuss free radicals and their effects on our cellular systems in detail.

Valence Electrons and Bonding

Understanding chemical bonds is a fundamental concept in grasping chemistry. In 1916, the Lewis theory of bonding introduced by Gilbert Newton Lewis simplified the observations of the time by combining theories together and providing a visual representation of what these bonds actually looked like. The Lewis theory focused on the importance of valence electrons and the octet rule.

"Octet rule states that in forming compounds, atoms gain, lose or share electrons to give a stable electron configuration characterized by eight valence electrons. This rule is applied to the main-group elements of the second period" ("Lewis Formulas and Octet Rule"). Valence electrons are electrons on the outermost shell of an atom that make chemical bonding possible. There are some exceptions to this rule—bonds that contain an odd number of electrons, bonds that do not have enough electrons to complete an octet, and atoms that fill shell level three or higher all fall outside of the guidelines.

In a covalent bond, the participating atoms both contribute a valence electron to form a shared pair. These are the types of bonds that form sugars like glucose and fructose. A network covalent bond is one where the same structure is replicated over and over, forming very large molecules. A diamond is an example of a network covalent bond where carbon atoms bond to themselves. Metallic bonds are interesting. In pure metals, the valence electrons are not firmly attached to any specific atom. They are delocalized due to their low ionization energy; this means that they do not belong to any specific region. Because of this, nearby atoms tend to share these loosely bound electrons, and we call this interaction bonding. The more electrons participate in the sharing, the stronger the bond. An ionic bond is formed when there is an electrostatic attraction between oppositely charged ions. These ions represent atoms that have either lost electrons (cations) or gained electrons (anions). Note that exclusively ionic bonds where one atom grabs an electron from another do not exist; there is always some involvement of electron sharing. The term ionic bond is used when the electronegativity (the chemical property of attraction) between the atoms is stronger than the equal sharing of electrons. Ionic bonds are used in compounds within the body that determine and hold the shape of proteins and chromosomes through the attraction of certain atoms. Ionic bonding also affects muscle contraction. As the positive and negative charges move throughout the cell, buildups occur on the inside and outside of the cell, which triggers the muscle to contract.

Bonds that are partially ionic and covalent are polar covalent bonds. Hydrogen bonds, also known as hydrogen bridges, are situations where a hydrogen atom is found between two other atoms whose electrons are eager to bond. These bonds are weaker than ionic and covalent bonds and usually occur with oxygen, fluorine, or nitrogen. Hydrogen bonds are what hold our DNA together.

Some atoms, such as the noble gases, contain closed shells or filled shells and are chemically stable. These do not often undergo chemical reactions. "Atoms with one or two more valence electrons than are needed for a "closed" shell are highly reactive because the extra valence electrons are easily removed to form a positive ion. Atoms with one or two valence electrons fewer than are needed to form a closed shell are also highly reactive because of a tendency either to gain the missing valence electrons (thereby forming a negative ion), or to share valence electrons (thereby forming a covalent bond)" (Sanghera 2011). The stability and omnipresence of atoms undeniably relies on their ability to form bonds.

Molecules and Compounds

Molecules are groups of two or more atoms that are electrically neutral and held together through chemical bonds. Within a molecule, electrons move under the influence of several nuclei and occupy molecular orbitals much as they can occupy atomic orbitals in isolated atoms (Trefil 2015). A molecule may be homonuclear, consisting of the same element, such as H_2 (hydrogen gas), or a chemical compound, which involves two or more different elements like CO_2 (carbon dioxide). Molecules are common in organic matter (substances containing carbon), our oceans, and our atmosphere. The majority of the solid composite materials on earth, such as the crust, mantle, core, and so forth, are made of minerals that form chemical compounds but have molecules that are not identifiable!

There are also ionized crystals (salts), covalent crystals, gases and metallic solids that exhibit the same characteristics of chemical bonding with unidentifiable molecules. Molecules have fixed equilibrium geometry; this means that each molecule exists in a standard shape to conserve its energy. Their angles and the lengths of their bonds are in a perpetual state of oscillation due to their rotations and vibrations. The structure and chemical formula of a molecule are the main characteristics that determine its properties and, most importantly, its reactivity.

In chemistry, the fundamental principle is that atoms from varying elements can bind together to form new chemical compounds. Remember that the main difference between a molecule and a compound is that a molecule is made of two or more atoms; a compound is made of two or more different elements. Essentially, not all molecules can be classified as compounds, but all compounds are molecules. "Methane, for example, which is formed from the elements carbon and hydrogen in the ratio four hydrogen atoms for each carbon atom, is known to contain distinct CH_4 molecules. The formula of a compound—such as CH_4—indicates the types of atoms present, with subscripts representing the relative numbers of atoms (although the numeral 1 is never written)" (Norman 2014).

Chemical compounds grant us an astonishing array of properties and capabilities. At regular temperatures, they may appear in gas, solid, or liquid form. These compounds can be extremely beautiful, spanning the colors of the rainbow; entirely toxic to human exposure and consumption; or absolutely necessary to the survival of life-forms—the possibilities are endless. To say that chemical compounds are sensitive to change would be an understatement; the substitution of a single atom can alter the entire composition of the compound. The color, levels of toxicity, structure, and even the smell of the compound might be subject to change.

Compounds are classified as either organic or inorganic. Many of the organic compounds were originally separated from living organisms—hence the name. Typically, they contain carbon atoms, and because of the multiple ways carbon atoms can bond with one another and other elements, there are millions of organic compounds. Those that are not organic compounds are simply specified as inorganic compounds. "Within the broad classifications of organic and inorganic are many subclasses, mainly based on the specific elements or groups of elements that are present. For example, among the inorganic compounds, oxides contain O^{2-} ions or oxygen atoms, hydrides contain H^- ions or hydrogen atoms, sulfides contain S^{2-} ions, and so forth. Subclasses of organic compounds include alcohols (which contain the -OH group), carboxylic acids (characterized by the -COOH group), amines (which have an $-NH_2$ group), and so on" (Norman 2014).

Up until this point, we have examined the discovery, development, structure, and application of atoms, molecules, and their elementary particles in an isolated discussion. For our purposes, we need to know how the environmental and biological factors of aging affect atoms in the human body. Now that we have some background knowledge about what these building blocks are, let us go further and take a look inside the human body!

Composition

From an elemental perspective, the human body is composed of six main chemicals (by mass) followed by multiple trace elements. The most abundant element, supplying 65 percent of body mass, is oxygen (O). Approximately 75 percent of the body is water, and nine-tenths of the mass that makes up water is oxygen. Oxygen is also used in respiration, providing nutrients for the body. Carbon (C) makes up 18 percent of the body. All organic molecules contain carbon. "Carbon also is found as carbon dioxide or CO_2. You inhale air that contains about 20% oxygen. Air you exhale contains much less oxygen, but is rich in carbon dioxide" (Helmenstine 2014).

Hydrogen (H) makes up 10 percent of the human body. Most of the hydrogen exists in the form of water. Some of its functions are to moisten and lubricate joints, to aid in digestion and the transportation of waste, to regulate healthy body temperatures, and many more! "The H^+ ion can be used as a hydrogen ion or proton pump to produce ATP and regulate numerous chemical reactions. All organic molecules contain hydrogen in addition to carbon" (Helmenstine 2014). Nitrogen (N) accounts for 3 percent of the human body. Organic molecules contain nitrogen, and it can also be found in the lungs, since it is the fundamental component found in the air we breathe. Calcium (Ca), found primarily in your bones, makes up 1.5 percent of the human body. It provides the rigidness and strength needed to support the body and is also found in our teeth and is an important part of muscle function. Phosphorus (P) makes up 1.5 percent of the body. "Potassium is an important mineral in all cells. It functions as an electrolyte and is particularly important for conducting electrical impulses and for muscle contraction" (Helmenstine 2014). The rest of the elements are found only in traces and are as follows: sulfur (S), sodium (Na), chlorine (Cl), magnesium (Mg), boron (B), chromium (Cr), cobalt (Co), copper (Cu), iodine (I), fluorine (F), iron (Fe), magnesium (Mn), selenium (Se), silicone (Si), tin (Sn), vanadium (V), molybdenum (Mo), and zinc (Zn). These different elements make up the chemistry that is you! But it doesn't stop there; the picture gets even bigger. The body has a way of organizing its components in a way that is self-sufficient and productively supportive.

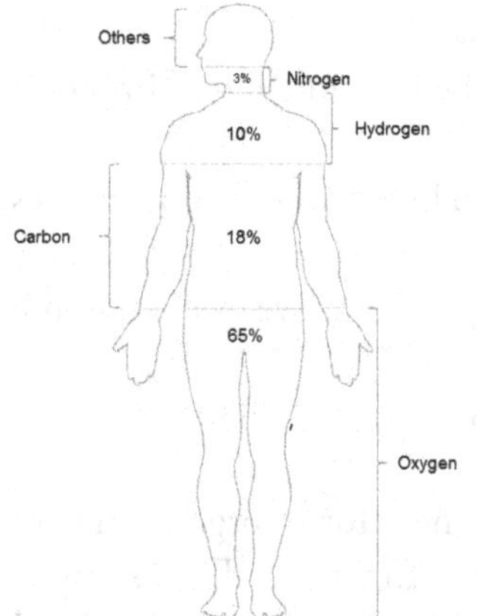

Element	Symbol	% in Body
Oxygen	O	65.0
Carbon	C	18.5
Hydrogen	H	9.5
Nitrogen	N	3,2
Calcium	Ca	1.5
Phosphorous	P	1.0
Potassium	K	0.4
Sulfur	S	0.3
Sodium	Na	0.2
Chlorine	Cl	0.2
Magnesium	Mg	0.1
Trace elements include: boron (B), chromium (Cr), cobalt (Co), copper (Cu), fluorine (F), iodine (I), iron (Fe), manganese (Mn), molybdenum (Mo), selenium (Se), silicon (Si), tin (Sn), vanadium (V) & zinc (Zn)		Less than 1.0

The human body in constructed through different levels of function and structure. The smallest part of this composition is the atom. Groups of atoms create molecules that create cells, groups of cells make up tissue, tissue makes up organs, and organs form organ systems! It's a chain reaction of responsibility.

The cell conducts varying duties of great importance to the body system that keep us running without conscious effort. For instance, your cells are responsible for the exchange of oxygen and carbon dioxide when you breathe, and your cells are responsible for providing the enzymes that speed up the breakdown of your food to convert it into useable energy during digestion. But it doesn't stop there; each cell performs the same tasks individually that your body does as a whole—reproduction, energy conversions, food digestion, waste disposal, and taking in oxygen. The atoms and molecules that make up these cells are all specialized. This means that they have distinct functions. There are many different types of cells, such as blood, nerve, and muscle cells. When these cells band and operate together, they form tissue.

There are four basic types of bodily tissues:

- Nerve tissue is responsible for the transmission of impulses and the formation of nerves.
- Connective tissue, found both in blood and in bones, serves to support and bind body parts.
- Muscle tissue allows your body parts to move by contracting and relaxing.
- Epithelial tissue lines organs and covers the body.

When these tissues work together, they form organs. These organs perform specialized physiological functions. For example, the stomach contains all four classes of tissue and is the localized point of the digestion of food. An organ system is a combination of different organs that work together. There are eleven different types of organ systems:

- The muscular system provides movement and support. Also, due to the high metabolic rate, it produces heat.
- The integumentary system includes skin, subcutaneous fat, hair, and nails. This system protects us from the environment, aids in regulating temperatures, and gives us characteristics that physically identify individuals.
- The skeletal system, which includes bones, cartilage, ligaments, joints, and tendons, provides protection for our organs and structural support in conjuncture with the muscular system. It also operates as a storage facility for minerals, such as magnesium, calcium, and phosphorus.
- The circulatory system—the heart, blood vessels, and blood—delivers oxygen to every cell, tissue, and organ in the body. It also transports electrolytes, immune cells, hormones, and other important substances throughout the body.
- The lymphatic system includes lymphatic vessels and lymph nodes. Lymph is a clear fluid that your lymphatic system carries from tissues and organs back to the heart. This

small amount of liquid is the product of leaky blood vessels that secrete into nearby tissue. This system also transports absorbed fat and immune cells.

- The respiratory system—the nose, trachea, mouth, airways, lungs, and diaphragm—is responsible for the absorption of oxygen and the release of carbon dioxide in the bloodstream. It is also involved in vocalization, the ability to speak and sing.
- The endocrine system is composed of the adrenals, thyroid, pancreas, pituitary, hypothalamus, gonads, and pineal glands. All hormone-producing organs and glands are part of the endocrine system. Hormones regulate functions and activities in the body.
- The urinary/excretory system involves the kidney, liver, bladder, urethra, and large intestine. This system's function is to remove unnecessary and excessive waste and bodily fluids, which might otherwise become toxic, from the body.
- The reproductive system includes Cowper's gland, Bartholin's gland, the clitoris, the penis, the ovaries, the testes, the prostate gland, Skene's gland, and the mammary glands. The human reproductive system functions as a system of fertilization through sexual intercourse that results in childbirth. Hormones in the female body produces an ova (an egg cell) every twenty-eight days. If the egg is fertilized, a fetus will develop.
- The mouth, throat, esophagus, stomach, small intestine, and large intestine form the digestive system. This system takes on the complicated process of turning the food we eat into energy, creating waste, and then disposing of it. The body uses this energy to grow and repair the cells needed for survival.

The organ systems are not completely exclusive; some organs are involved in more than one system, and some systems work together. The body as a whole operates as a single organ!

So what is the connection between atoms and the human body? We know that these particles make up our cells, which form these complex systems that orchestrate our physical bodies, but why is this important? What is the application? Cellular health is the key component to our physical health; our bodies are dependent on it! In order for our bodies to function at optimal capacity we need proper nutrition, effective management of stress, and an environment free of toxins and poisons. Preventative medicine as part of a healthy lifestyle is a serious and promising defense against disease. Nutrition on a cellular level aids all of these; assuring that the most fundamental processes—nutrients entering the cell and waste leaving the cell—are properly managed. Each cell has a membrane that controls what enters and exits the cell. "These functions are largely governed by the health and proper functioning of cell membranes which cover every cell and which are composed largely of fats, lipids and sterols. In order to absorb these essential lipids and sterols, they have to be provided in adequate amounts in the diet and have to be digested and absorbed into the blood where they can provide optimum cellular nutrition to every part of the body" (ANC 2015).

Aging is a natural process that is the by-product of many entangled factors. The most basic way to understand this progressive and steady cellular deterioration is as the body's inability to effectively replicate its cells. "The average adult body is composed of around 30 trillion cells, each of them attempting to produce enough energy to serve their two primary functions: 1. Perform its specific function within the body. 2. Regenerate itself (and keep the DNA safe from damage)" (Hart 2015).

Our cells are engineered specifically to process certain nutrients. These nutrients provide the spark needed to create and sustain energy, repair damage, and keep the cellular regeneration process going. If a cell is unable to perform these functions, it becomes problematic and offsets the system. Proper nutrition and detoxing are good ways of setting balance to improve and maintain cellular health.

A Deeper Look Into How Our Cells Work

The mitochondria in the cell function as the powerhouse; they're almost like a miniature digestive system. The surrounding membrane is responsible for turning nutrients into energy and managing metabolism. The process as a whole is called cellular respiration. As the cell takes in nutrients, the reaction uses oxygen. This process is what causes free radicals within the body. If you recall from earlier, free radicals are highly reactive, positively charged, unpaired electrons. They are produced not only through metabolic practices but also as a response to environmental toxins. An unpaired electron causes instability and damage to the body. These free radicals are destructive and messy. They desperately seek to become stable and paired. In these attempts, they thrash around until they successfully steal an electron from a healthy cell. The robbed cell then becomes a free radical. This chain reaction of electron stealing results in a multitude of damaged cells.

"Free radical imbalance within the body results in oxidative stress, leading to cellular damage. Once the negative feedback loop of oxidative stress is established in the body, cellular deterioration and organs malfunction start affecting the entire body. This imbalance can lead to cross-linking of atomic structures and DNA in the body, damaging cellular DNA, respiratory chain, membranes and proteins. DNA cross-linking can in turn lead to various increasing cellular DNA defects, fueling aging, skin wrinkling, cancer, heart diseases and degenerative diseases" (Hart 2015). In the next chapter, we will discuss the free radical theory of aging, what it is, how it affects the body, and what can be done about it!

The ozone layer may be the beginning and the end of mortals.

—Alfred Sparman, MD

4

Free Radicals and the Theory of Aging

A free radical is classified as an unstable, short-lived atom or molecule that lacks a balanced pairing of electrons. There is no innate or implied charge when it comes to these radicals; they may be positive, negative, or neutral. In this chapter, we will be addressing specifically the free radical subgroup known as reactive oxygen species (ROS). These are neutrally charged molecules that contain oxygen whose outer orbits have one or more electrons that are not paired. A free radical forms through homolytic bond cleavage. This occurs when the introduction of some energy, such as heat, breaks a covalent bond.

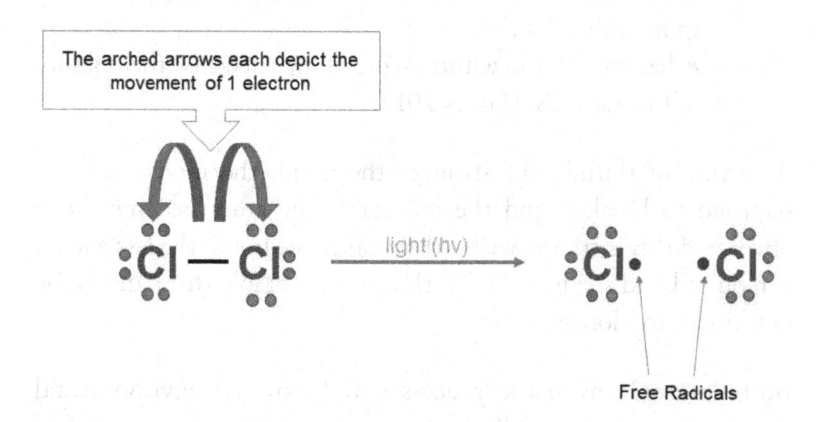

When this happens, the shared electron pair between the two atoms is split evenly; each atom takes back its electron. This offset, this oddly numbered arrangement of electrons, causes the atom to become unstable and highly reactive. To say that a chemical is reactive simply means that when it comes into contact with another chemical, there is some detectable change or effect. In other words, there is a reaction. If something is highly reactive, it does not take much stimulus to cause a reaction within that substance. The electron imbalance within a free radical causes it to seek stability; this is what makes it reactive. The free radical looks to steal the necessary electrons from neighboring molecules, resulting in a domino effect of lost stability. However, not all free radicals are extremely unstable. The strength of the covalent bond formed between the molecule's radicals depends on their level of stability.

- Stability increases in the order methyl »»primary »» secondary »» tertiary.
- Free radicals are stabilized by resonance.
- Free radicals are stabilized by adjacent atoms with lone pairs.
- Free radicals increase in stability as the electronegativity of the atom decreases.
- Free radicals increase in stability as we go down the periodic table (that is, as atoms increase in size).
- Free radicals decrease in stability as we go from sp^3 to sp^2 to sp hybridization.
- Adjacent electron withdrawing groups decrease the stability of free radicals. (James 2013)

As a rule of thumb, the stronger the bond, the more energy is required to break it and the less stable the atom becomes. The obverse of this is true as well—the weaker the bond, the less energy is required and the more stable the atom is. Stable free radicals, by definition, live longer.

Both internal enzymatic processes and external environmental forces can create free radicals. Some environmental causes of free radical pollutants are cigarette smoking and secondhand smoke, air pollution, industrial chemicals, radiation from the sun, processed foods, prescription and recreational drugs, and ozone depletion.

> The pollutants produced by modern technologies often generate free radicals in the body. The food most of us buy contains farm chemicals, including fertilizers and pesticides that produce free radicals when we ingest them. Prescription drugs often have the same effect; their harmful side-effects may be caused by the free radicals they generate.
>
> Processed foods frequently contain high levels of lipid peroxides, which produce free radicals that damage the cardiovascular system. Cigarette smoke generates high free-radical concentrations; much of the lung damage associated with smoking is caused by free radicals. Air pollution has similar effects. Alcohol is a potent generator of free radicals (although red wine contains antioxidants that counteract this effect). In addition, free radicals can result from all types of electromagnetic radiation-including sunlight. Exposure to sunlight generates free radicals that age the skin, causing roughness and wrinkles. (Sharma et al. 1998)

Stress, both emotional and physical, can cause free radical buildup as well. Studies show that exercise can also create free radicals. Acute physical exertion causes a buildup of free oxygen radicals due to the excessive intake of oxygen and the initiation of certain metabolic processes. However, antioxidants are also produced during exercise, so the two naturally cancel each other out when exercise is conducted in moderation.

Within the cell, the main formation of free radicals occurs during the metabolic processes of the mitochondria that turn food into chemical energy; this causes radicals to be produced naturally as a by-product.

Cellular respiration, also known as the electron transport chain, is the act of releasing energy through electron transport and proton pumping while simultaneously conserving energy to produce ATP.

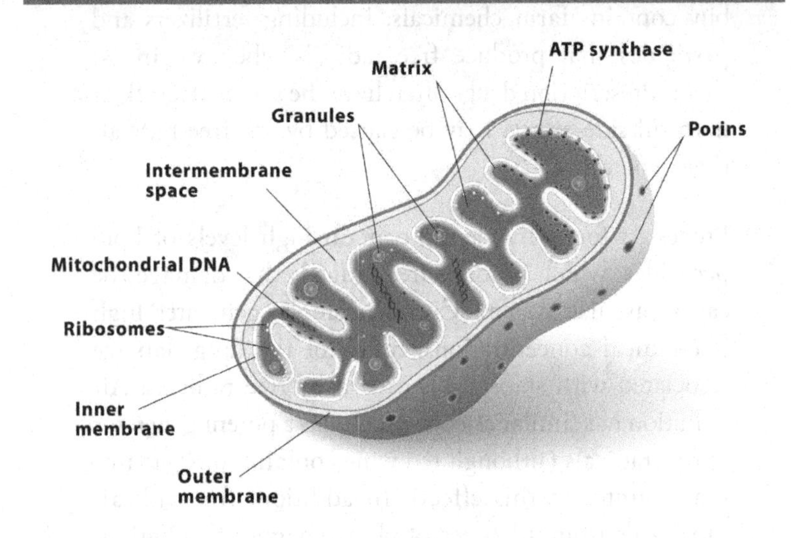

During the transport, a sequence of compounds transfer electrons between electron donors and electron acceptors through a process called redox. When an electron is gained, there is a decrease in the oxidation of the atom or molecule (in chemistry, this is called a reduction); when an electron is lost, there is an increase in oxidation. These transfers release the energy needed to produce protons that accompany the electrons across the cell's membrane. As a result, there is a buildup of both negative and positive charges that produces an electrochemical reaction. This well of energy is called the ATP synthase and creates a thermodynamic state of work potential to turn ADP, an important compound in the metabolic process (a storage molecule), into ATP, a coenzyme that can be described as molecular currency, through a process called oxidative phosphorylation in which a protein enzyme is turned on or off by the addition of a phosphate. ATP acts as a recyclable molecular transfer of energy that fuels cells chemically whenever and wherever energy is needed. For example, physical movements, such as muscle contractions, require ATP. However, although this transport chain is highly effective in sustaining cellular function, there is tendency for a portion of the electrons to leak back into the mitochondria through the redox centers (complexes I, II, III, and IV) causing the formation of ROS. "In the mitochondrial respiratory chain, Complex IV (cytochrome oxidase) retains all partially reduced intermediates until full reduction is achieved. Other redox centers in the electron transport chain, however, may leak electrons to oxygen, partially reducing this molecule to superoxide anion (O2−•). Even though O2−• is not a strong oxidant, it is a precursor of most other reactive oxygen species, and it also becomes involved in the propagation of oxidative chain reactions" (Turrens 2004).

The metabolic rate for increased production of the superoxide anion is dependent on two factors: the slowing down of electrons in the transport chain, which results in a surplus of electron donors, and the increase in concentration of oxygen. Electrons are slowed down during the oxidative phosphorylation of ADP. Once ADP is no longer present, the protons in the ATP synthase halt, and there is a stagnant buildup of H+ in the gradient. Without that extra pump, there is a resulting change in speed of passing electrons. The respiratory chain now has an increase in reduction. Ultimately, the destination for all electrons in the chain is an oxygen molecule. Once the chain is complete, the oxygen molecule will reduce into water. However, in approximately 2 percent of cases, oxygen reduces prematurely and incompletely and subsequently becomes a superoxide anion, a free radical.

It is important to bear in mind that free radicals are not innately problematic; in fact, some are useful in defending against pathogens or harmful microbes! "The body tries to harness the destructive power of the most dangerous free radicals—the oxy radicals and ROS—for use in the immune system and in inflammatory reactions. Certain cells in these systems engulf bacteria or viruses, take up oxygen molecules from the bloodstream, remove an electron to create a flood of oxy radicals and ROS, and bombard the invader with the resulting toxic shower. This aggressive use of toxic oxygen species is remarkably effective in protecting the body against infectious organisms" (Sharma et al. 1998).

Problems occur when there are too many free radicals in the body. Oxidation can cause biological damage to a number of prominent chemical classes, including lipids, nucleic acids (like DNA), proteins, carbohydrates, and amino acids. Oxidative stress develops when the production of free radicals exceeds the body's ability to detoxify and neutralize the harmful effects with antioxidants. These radicals can also interfere with normal cellular signaling and act as their own messengers in redox signaling. In normal, healthy cells, signaling is an intracellular form of communication and coordination that allows neighboring cells to be aware of and respond to changes within their environment. Signaling regulates the processes within the cell. For example, in the mitochondria, a signal may tell how much ATP needs to be produced and when it will be needed. Signaling cells release the signaling molecules into the bloodstream to targeted receptor cells, and upon arrival they are either absorbed by that cell through a process called endocytosis or enter through its membrane. A signaling molecule can function in a number of ways; the three main types are endocrine, paracrine, and autocrine signaling. These molecules can be identified chemically as lipids, proteins, or amino acids, among other classifications. In endocrine signaling, hormones are released to manage secretions. "Hormones can be: small lipophilic molecules that diffuse through the cell membrane to reach cytosolic or nuclear receptors. Examples are progesterone and testosterone, as well as thyroid hormones. They generally regulate transcription; or water soluble molecules that bind to receptors on the plasma membrane. They are either proteins like insulin and glucagons, or small, charged molecules like histamine and epinephrine" (Nogales 2008).

In paracrine signaling, the signaling molecules are from the nervous system. "The signaling molecule affects only target cells in the proximity of the signaling cell. An example is the conduction of an electric signal from one nerve cell to another or to a muscle cell. In this case, the signaling molecule is a neurotransmitter" (Nogales 2008).

Autocrine signaling cells are called cytokines, and they are emitted from the immune system. "In autocrine signaling cells respond to molecules they produce themselves. Examples include many growth factors. Prostaglandins, lipophilic hormones that bind to membrane receptors, are often used in paracrine and autocrine signaling. They generally modulate the effect of other hormones. Once a signaling molecule binds to its receptor it causes a conformational change in it that results in a cellular response" (Nogales 2008).

Other types of cellular signaling are juxtacrine and intracrine. The former targets nearby cells and is transmitted through lipids or proteins. The latter form of signaling produces its own signal and then receives it. So what does this all mean? The purposes of cellular signaling are to maintain the integrity of cells and communicate functions effectively. In redox signaling, the cell knows that it's damaged and sends off messages to the tissue that it will either try to repair itself or undergo cell death.

The following is a table of common oxidants:

Oxidant	**Description**
$\bullet O^{2-}$, superoxide anion	One-electron reduction state of O_2 formed in many autoxidation reactions and by the electron transport chain. Rather unreactive but can release Fe^{2+} from iron-sulfur proteins and ferritin. Undergoes dismutation to form H_2O_2 spontaneously or by enzymatic catalysis and is a precursor for metal-catalyzed $\bullet OH$ formation.

H_2O_2, hydrogen peroxide	Two-electron reduction state formed by dismutation of $\bullet O^{2-}$ or by direct reduction of O_2. Lipid soluble and thus able to diffuse across membranes.
•OH, hydroxyl radical	Three-electron reduction state formed by Fenton reaction and decomposition of peroxynitrite. Extremely reactive and will attack most cellular components.
ROOH, organic hydroperoxide	Formed by radical reactions with cellular components, such as lipids and nucleobases.
RO•, alkoxy and ROO•, peroxy radicals	Oxygen-centered organic radicals. Lipid forms participate in lipid peroxidation reactions. Produced in the presence of oxygen by radical addition to double bonds or hydrogen abstraction.
HOCl, hypochlorous acid	Formed from H_2O_2 by myeloperoxidase. Lipid soluble and highly reactive. Will readily oxidize protein constituents including thiol groups, amino groups, and methionine.
ONOO-, peroxynitrite	Formed in a rapid reaction between $\bullet O^{2'}$ and NO•. Lipid soluble and similar in reactivity to hypochlorous acid. Protonation forms peroxynitrous acid, which can undergo homolytic cleavage to form hydroxyl radical and nitrogen dioxide.

(Mandal 2010)

The severity of damage to the cell depends on its ability to recover and return to its original state. Minor discrepancies can be reversed; however, there are cases where cell death is inevitable. Apoptosis is a purposeful and controlled type of programmed cellular death. Cellular systems have to regulate order in their production (cell division) and in their destruction.

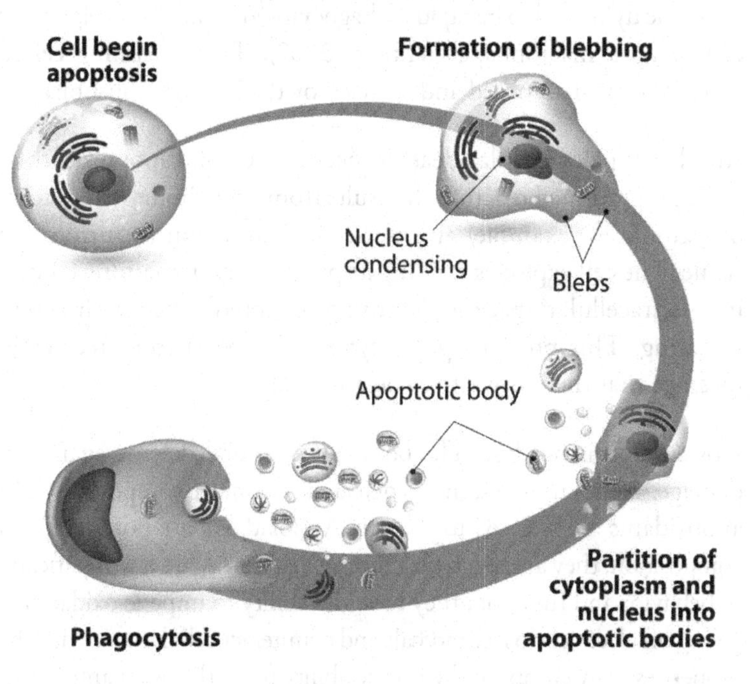

If a cell is no longer needed or shows signs of damage, it activates its intercellular program for deletion. For example, in human embryological development, our fingers and toes start off shaped somewhat like paws. Through apoptosis, the excess tissue dies off and leaves five individually separated digits. Furthermore, "In the developing vertebrate nervous system up to half or more of the nerve cells normally die soon after they are formed. In a healthy adult human, billions of cells die in the bone marrow and intestine every hour" (Alberts 2002).

Although out of context this mechanism may seem extreme, the key thing to remember is that apoptosis is about balancing the rate of cellular production. Too much, and there would be shrinkage in the tissue; too little, and there would be an overgrowth. "A cell that undergoes apoptosis dies neatly, without damaging its neighbors. The cell shrinks and condenses. The cytoskeleton collapses, the nuclear envelope disassembles, and the nuclear DNA breaks up into fragments. Most importantly, the cell surface is altered, displaying properties that cause the dying cell to be rapidly phagocytosed, either by a neighboring cell or by a macrophage" (Alberts 2002). The remaining cellular material is then absorbed and recycled by the cell that consumes it.

Another type of cellular death is necrosis. Unlike apoptosis, this is not a programmed reaction. It results from acute injury (from toxins or trauma, for example) and causes inflammation to surrounding tissue. The cell explodes, spilling its proteins and remaining contents into extracellular space and causing an interference with cellular signaling. This prevents phagocytes (cell eaters) from receiving a message that there is waste to be removed.

However, all is not lost. The body has a natural defense against cell damage. As soon as trauma is detected, chemical compounds called antioxidants are released to the site. Antioxidants are not identical in composition; they are not a substance. What they share is a classification of behavior and function; they have the ability to impede oxidation by giving off electrons to free radicals and counteract cell damage with their properties. Any compound with the ability to do this is an antioxidant. The reason antioxidants are able to react so remarkably is that they themselves become oxidized in the process. As they lose electrons, they become reducing agents, and their oxidative state is usually lowered.

The body absorbs antioxidants naturally from fruits and vegetables, but there are also artificial supplements available that one can take in order to meet dietary requirements. These redox compounds can be found in blueberries, blackberries, green tea, dark chocolate, grapes, garlic, onion, ground turmeric, ginger root, ground cinnamon, and the list goes on and on. There are thousands of compounds that contain antioxidant properties, but the most common in our everyday diet are lycopene; beta-carotene; and vitamins A, C, and E. Antioxidants are also found in foods that contain the element selenium and in flavonoids as well as in lignans, such as nuts, red meat, shellfish, red wine, soy, barley, and rye.

Antioxidants' performance varies. We can categorized antioxidants by their kinetics or mechanisms of action, atomic structure, and solubility. The chart below offers some insight into how different types of antioxidants function.

Alphabetical Name	Antioxidant Category	Examples
Antioxidant C	Carotenoids beta-carotene and lycopene	Lutein
Antioxidant E	Enzymes, SOD	Catalase, GPx
Antioxidant G	Glutathione	Glutathione
Antioxidant H	Hormones	Melatonin, estrogen
Antioxidant L	Lipid-associated chemicals	Ubiquinol-10, N-acetyl cysteine, lipoic acid
Antioxidant M	Minerals	Zinc, selenium, copper
Antioxidant P	Phenolics	Quercetin, catechin
Antioxidant S	Saponines, steroids	Cortisone, estradiol, estriol
Antioxidant V	Vitamins alpha-tocopherol	Ascorbic acid

(Flora 2009)

Kinetically, we can describe how an antioxidant moves and reacts, reduces, and so forth, using six classifications.

1. Antioxidants that break chains by reacting with peroxyl radicals that have weak O-H or N-H bonds: phenol, napthol, hydroquinone, aromatic amines, and aminophenols.

2. Antioxidants that break chains by reacting with alkyl radicals: quinones, nitrones, and iminoquinone.

3. Hydroperoxide decomposing antioxidants: sulfide, phosphide, and thiophosphate.

4. Metal-deactivating antioxidants: diamines, hydroxyl acids, and bifunctional compounds.

5. Antioxidants that terminate cyclic chains: aromatic amines, nitroxyl radicals, and variable valence metal compounds.

6. Synergy of action of several antioxidants: phenol sulfide in which the phenolic group reacts with peroxyl radicals and sulfide group with hydroperoxide. (Flora 2009)

The structure of the antioxidant also affects the strength of its bond energy and reactivity. The active group, or the functional group, consists of the atoms or bonds within a molecule whose performance defines the specific characteristics and reactions of the molecule. For instance, with the free radical chemical compound DPPH (diphenyl-1-picrylhydrazy), the scavenging activity associated with H_2O_2 produces the following result: the phenol compounds (pertaining to hydroxyl group O-H) were more active than the aniline compounds (pertaining to phenyl group N-H) due to their low bonding energies. "The antioxidant activity related to the compound structure was found to be dependent on the number of the included active group (OH or NH_2). The more active the compound is, the more included active groups. The position of the active groups also plays an important role of structure–antioxidant relationship activity. The *ortho* position was found to be the more active one, due to its ability to form intramolecular hydrogen bonding (iHB), followed by *para* position and then *meta* position of compounds" (Ferreira 2006).

In ortho positions, two related active groups are fixed together in a ring of benzyne (a type of triple bond) in the first and second position. In the metaposition, this occurs in positions one and three, and in the paraposition, this occurs in positions one and four.

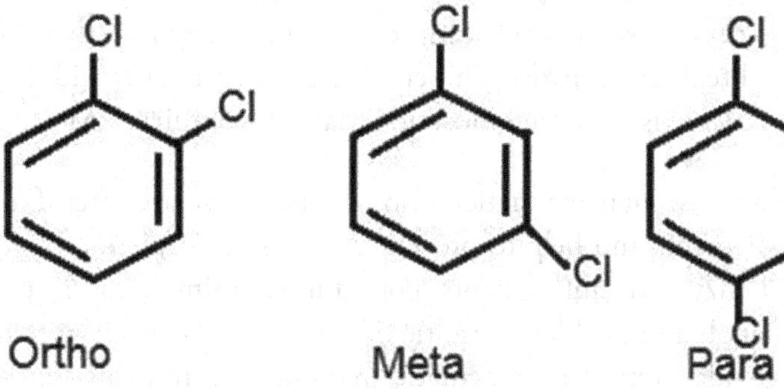

Cl Cl Cl Cl Cl Cl

Ortho Meta Para

The size of an antioxidant also contributes to its function. Small-molecule antioxidants work to clean up or scavenge reactive oxygen species; they transport them away through neutralization. Large-protein antioxidants are usually enzymatic; they swallow or absorb the reactive oxygen species to prevent any damage to important proteins.

Enzymatic antioxidants are extremely beneficial! The body produces them naturally, and they do the hard work of breaking down and deleting free radicals. With the help of trace metals like copper, zinc, and iron, these antioxidants conduct a process that can take dangerous oxidants and turn them into hydrogen peroxide and then convert them to water.

Mercola provides the following list of the main antioxidants in your body:

- **Superoxide dismutase (SOD)** can break down superoxide into hydrogen peroxide and oxygen, with the help of copper, zinc, manganese, and iron. It is found in almost all aerobic cells and extracellular fluids.
- **Catalase (CAT)** works by converting hydrogen peroxide into water and oxygen, using iron and manganese cofactors. It finishes up the detoxification process started by SOD.

- **Glutathione peroxidase (GSHpx)** and **glutathione reductase** are selenium-containing enzymes that help break down hydrogen peroxide and organic peroxides into alcohols. They are most abundant in your liver. (Mercola)

There are also nonenzymatic antioxidants. These are often found in supplements and help to prevent unnecessary depletion of your natural enzymatic antioxidants. Found in vitamins C and E, plant polyphenols, and GSH, these antioxidants inhibit free radical chains and provide support to enzymatic free radicals. You can consider them your first line of oxidative defense.

Moving on, the solubility of a given substance refers to its ability to dissolve in water or in another substance. Antioxidants fall into one of two categories based on their solubility: hydrophilic or hydrophobic. A hydrophilic antioxidant dissolves in water; a hydrophobic antioxidant dissolves in lipids. Both forms are necessary because of the architecture of our cells. The insides of cells and their surrounding fluids are composed mostly of water, whereas the cellular membrane is comprised almost entirely of fat. Free radicals are capable of attacking in either environment, so to enforce quality protection, you need both types of antioxidants working together. "Lipid-soluble antioxidants are the ones that protect your cell membranes from lipid peroxidation. They are mostly located in your cell membranes. Some examples of lipid-soluble antioxidants are vitamins A and E, carotenoids, and lipoic acid. Water-soluble antioxidants are found in aqueous fluids, like your blood and the fluids within and around your cells (cytosol or cytoplasmic matrix). Some examples of water-soluble antioxidants are vitamin C, polyphenols, and glutathione" (Mercola).

Historically, the term *antioxidant* referred exclusively to a chemical that inhibited oxygen consumption. "In the late 19th and early 20th centuries, extensive study concentrated on the use of antioxidants in important industrial processes, such as the prevention of the metal corrosion, the vulcanization of rubber, and the polymerization of fuels in the fouling of internal combustion engines" (Lobo et al. 2010). In the field of biology, antioxidants were seen as important primarily because they prevent food from going bad. The oxidation of unsaturated fats is what causes rancidity; antioxidants were used to prevent this. Their activity was discovered through simple calculations in which a fat was inserted into a tightly closed, oxygen-filled container. Researchers then measured the percentage of oxygen consumed over time. Oxidative rancidity of unsaturated fats produces free radicals that break their double bonds and produce a cleavage that releases harmful carbon compounds. It was then discovered that this entire process can be interrupted and halted with antioxidants! But what really revolutionized the field of biology and charted the course for understanding, utilizing, and declaring the major role that antioxidants play in living organisms was the classification of vitamins A (beta-carotene in its precursor form), C, and E as antioxidants. It was the discovery that substances with antioxidative properties were in fact oxidized themselves that led to further investigation of the possible mechanisms of action for antioxidants. Studies on how vitamin E impedes the continuation of lipid peroxidation paved the way for antioxidants to be classified as reducing agents that prevent the effects of oxidation.

Other important antioxidants that are produced by your body and are very important to your health are glutathione, ALA (alpha-lipoic acid), and CoQ10 (ubiquinone).

Glutathione, a tripeptide that's considered the most powerful antioxidant in the body, is found in every single cell. This intracellular antioxidant has the ability to amplify the effects of all other antioxidants in the fruits and vegetables that you consume as part of your daily diet. As a priority, glutathione works to protect your cells from both oxidative and peroxidative damage and environmental stresses. It also works to detoxify and flush out toxins, ensure proper use of energy, and prevent diseases commonly associated with aging.

ALA has the ability to suppress inflammation through gene expression, fortify insulin performance, and act as a powerful metal chelator (that is, it promotes molecular bonding of metal ions)—all as added bonuses to its free radical scavenging abilities. "ALA is the only antioxidant that can be easily transported into your brain, which offers numerous benefits for people with brain diseases, like Alzheimer's disease. ALA can also regenerate other antioxidants, like vitamins C and E and glutathione. This means that if your body has used up these antioxidants, ALA can help regenerate them" (Mercola).

CoQ10 provides natural protection from free radicals and is reduced by your body to ubiquinol to maximize its potential. It can be found in every cell and produces energy and provides support to your bodily systems. It also works to support heart health, maintain healthy blood pressure levels, and weaken the signs associated with aging.

We have spent an adequate amount of time describing the process and formation of free radicals and their counterparts, antioxidants. The rest of this chapter will focus on how the cellular damage implemented by free radicals results in disease and what role the theory of free radicals plays in our lives.

Free Radicals, Oxidative Stress, and Disease

As discussed earlier, an overabundance of free radicals results in the body undergoing oxidative stress. Metabolic or enzymatic reactions, environmental toxins, or a deficiency in antioxidants that results in damage to the cells or tissues all can cause oxidative stress.

Further, environmental factors can enhance the effects of enzymatic antioxidant deficiency and vice versa, so the initial cause of oxidative stress is not necessarily the only factor. Below is a list of diseases that have been linked to oxidative stress by ROS (adapted from Reuter et al. 2010):

- acute respiratory distress syndrome
- aging

- Alzheimer's
- atherosclerosis
- cancer
- cardiovascular disease
- diabetes
- inflammation
- inflammatory joint disease
- neurological disease
- obesity
- Parkinson's
- pulmonary fibrosis
- vascular disease
- rheumatoid arthritis

Whether oxidative stress is a direct cause of disease or a product of cell damage that results from the disease is unclear in many cases. However, the correlation between the two is clear, and the presence of excessive ROS seems to exacerbate disease. For example, let us take a look at the link between cardiovascular disease and oxidative stress. "Many disorders related to the metabolism of transition metals, amino acids or low molecular mass reductants are known to be connected with activities of antioxidant enzymes. Particularly, impairment in selenium uptake or synthesis of selenocysteine needed for Glutathione Peroxidases (GPx) may lead to deficiency and subsequent disorders such as cardiovascular ones" (Gospodaryov and Volodymyr 2012).

Heart failure occurs when there is deformity in the structure or function of the heart that interferes with its ability to receive or pump blood. The physical manifestation of this condition is the retention of fluid or a decrease in output flow. Heart failure is an increasing problem, especially in Western industrialized societies, and is a leading cause of death. With the graying or aging of the global population, it is a striking concern in public health, considering the link between heart failure and age. Ample experimental and clinical research supports the claim that oxidative stress is present in heart failure and aids in its development. "Excessive ROS causes cellular dysfunction, protein and lipid peroxidation, and DNA damage and can lead to irreversible cell damage and death, which have been implicated in a wide range of pathological cardiovascular conditions" (Tsutsui 2005; Tsutsui et al. 2011).

Oxidative stress reconfigures how the heart functions as heart failure continues to worsen. The role of ROS in heart failure is extremely important. These radicals corrupt proteins that are responsible for contraction by impairing their function, they interrupt regular cellular signaling and activate apoptosis, and they stimulate enzymes that cause the extracellular matrix of the heart to be remodeled. Once the level of ROS exceeds the body's capacity to neutralize antioxidants, oxidative stress attacks the complete biological integrity of the heart. "Oxidative stress has also been suggested as the major mechanisms causing endothelial (inner lining of blood vessels) dysfunction not only in atherosclerosis but also in Heart Failure (HF). Clinical studies suggested that endothelial dysfunction was independently associated with adverse long-term outcomes in patients with HF" (Tsutsui 2005; Tsutsui et al. 2011).

Experiments have studied the effects of oxidative stress on various young animals, plant forms, and even bacteria. In one case specifically, Hokkaido University Graduate School of Medicine used dogs and mice without any prior risk factors. The researchers induced one of two types of heart failure: heart failure following a myocardial infarction or rapid pacing–induced heart failure. As a result, the specimens showed structures similar to patients with heart failure. "The major biochemical markers of ROS generation, were elevated in the plasma and pericardial fluid of patients with HF and also positively correlated with its severity. The generation of •OH implies a pathophysiological significance of ROS in HF because •OH radicals are the predominant oxidant species causing cellular injury" (Tsutsui 2005; Tsutsui et al. 2011). This means that there was an undeniable increase in ROS within the myocardium.

It is important to note that a decrease in the production or impairment in the function of antioxidants can also cause a rise in ROS. A study conducted by Hill and Singal (Hill and Singal 1996) indicated that both the failing of antioxidants and the generation of oxidative stress contribute to heart failure and myocardial infarctions. These changes also have a direct impact on hemodynamic functions, blood flow, and circulation, further contributing to the theory of their involvement in pathological dysfunction.

However, in the study by Hokkaido, there was no evidence of decrease in behavior for enzymatic antioxidants! "GSHPx activity was even increased in the heart obtained from pacing-induced HF. Our results indicated that oxidative stress in HF might be primarily due to the enhancement of ROS generation rather than to the decline in antioxidant defense within the heart" (Tsutsui 2005; Tsutsui et al. 2011).

Inflammation is the body's healing defense against bacteria or other pathogens that cause cellular irritation. Physically, it may manifest as soreness, heat, swelling, or redness—all the symptoms we are familiar with from our childhood scrapes and tumbles. The process usually resolves quickly as blood vessels sprout, wounds scab, and skin recovers. However, in chronic inflammation, biological, physical, and chemical factors are linked to disease and cancers. Epidemiologists have long speculated about a connection between long-term inflammation and cancer, and the efficacy of anti-inflammatory therapies in both prevention and treatment has confirmed their speculation. "Rudolph Virchow first noted that inflammatory cells are present within tumors and that tumors arise at sites of chronic inflammation. This inflammation is now regarded as a 'secret killer' for diseases such as cancer. For example, inflammatory bowel diseases such as Crohn's disease and ulcerative colitis are associated with increased risk of colon adenocarcinoma and chronic pancreatitis is related to an increased rate of pancreatic cancer" (Beckman and Ames 1998).

The specific mechanisms by which chronic inflammation becomes cancer are still undergoing research; possible connections include instability and alterations in the genome and failure to activate cell death. What we do know, however, is that oxidative stress develops as a result of the necessary increase in oxygen and consequent accumulation of ROS to accompany the mast and leukocyte immunity cells to the site of inflammation. Over a prolonged period, this oxidized environment can become hazardous to neighboring cells and may cause once-normal healthy cells to morph into cancer cells through a process called carcinogenesis.

> Cancer is a multistage process defined by at least three stages: initiation, promotion, and progression. Oxidative stress interacts with all three stages of this process. During the initiation stage, ROS may produce DNA damage by introducing gene mutations and structural alterations of the DNA. In the promotion stage, ROS can contribute to abnormal gene expression, blockage of cell to cell communication, and modification of second messenger systems, thus resulting in an increase of cell proliferation or a decrease in apoptosis of the initiated cell population. Finally, oxidative stress may also participate in the progression stage of the cancer process by adding further DNA alterations to the initiated cell population. (Beckman and Ames 1998)

Recent studies have begun to provide evidence that links oxidative stress to tumor formation. RNS (reactive nitrogen species), which are derived from superoxide and nitric oxide, for example, are known to be a key participant in the proliferation of inflammation-related carcinogenesis through the activation of redox-sensitive transcription factors.

Chronic inflammation of the lungs, also known as asthma, is a disease that limits and remodels the pathways of airflow. Sufficient evidence has been provided to confirm that the imbalance between reducing and oxidation agents results in a more oxidized state. Both ROS and RNS are major players in the inflammatory reactions of the airways and act as determinants of the severity of the disease. An additional characteristic of asthma is a decrease in enzymatic antioxidant defense, which more than likely aids in the sustainment of inflammation. The ROS and RNS generated by inflammation attack lipids and proteins and the instant formation of O_2 radicals are the main causes of asthma attacks. The dominant nitrogen species found within the lungs is NO. "Autoxidation of NO with oxygen results in the formation of nitrite, a substrate for eosinophil peroxidase (EPO) and myeloperoxidase (MPO). Nitric oxide reacts with superoxide to form $ONOO^-$, which can nitrate tyrosine residues and thus damage enzymes, and structural and functional proteins. Higher NO levels are associated with higher risk of asthma, asthma severity, and greater response to bronchodilator agents" (Sahiner 2011).

The severity is also dependent on the production of neutrophils and eosinophils (white blood cells) by ROS that become reactive within the airways. Their offspring, a product called 3-bromotyrisine, is found to be much higher in the bronchoalveolar lavage fluid of those with asthma than that of those without. Environmental ROS also affect the respiratory system. Consider, for example, cigarette smoke: "Cigarette smoke is related to asthma exacerbations, especially in young children, and there is a dose-dependent relationship between exposure to cigarette smoke and rates of asthma. Cigarette smoke is a highly complex mixture of more than 4000 chemical compounds that are distributed in aqueous, gas, and the tar phase of the smoke. In the gas phase, the smoke contains high concentrations of O_2 and nitric oxide. They immediately react to form highly reactive peroxynitrite" (Sahiner 2011).

As mentioned earlier, the balance between ROS, RNS, and antioxidants in asthma patients is significantly impaired. Superoxide dismutase (SOD) is extremely important in that it counteracts superoxide, the most prevalent source of ROS production. The weakened activity of this enzyme in the lungs leaves cells vulnerable for oxidation. "Studies in large populations showed that the airway reactivity is inversely related to SOD activity. Transgenic mice that over-express SOD had decreased allergen-induced physiologic changes in the airways compared with controls. It seems that the lower SOD activity is partly a consequence of the increased oxidative and nitrative stress in the asthmatic airway and serves as a sensitive marker of airway redox and asthma severity" (Sahiner 2011).

Acute ischemic stroke is a neurological dysfunction or brain attack. It is the more common form of stroke in comparison to hemorrhagic stroke. The former is caused by a blocked blood vessel in the brain, and the latter is caused by a ruptured blood vessel that leaks into the brain. Like any other organ in the body, your brain depends on a constant supply of oxygenated blood to function properly. When this process becomes interrupted, brain damage ensues, and the results can be life altering. Studies have provided evidence that ischemic stroke, with or without restored blood flow, shows a positive correlation with the increased output of free radicals in mice. Human stroke patients show increased oxidative stress. This means that ischemic stroke when propagated by ROS damage induced by lipid, protein, and DNA oxidation may lead to cell toxicity and ultimately death without the ability to self-repair.

> Both neuronal necrosis (Patel et al. 1996) and apoptosis can be induced by oxidative stress. In addition to directly reacting with cellular molecules, there is increasing evidence that free radicals can act also by redox sensitive signal transduction pathways (Chan 2001). Experimental studies suggests that free radicals can induce cytochrome C release from mitochondria, which is an important step in the induction of apoptosis (Fujimura et al. 1999; Kim et al. 2000). Moreover the contribution of oxidative stress to apoptotic cell death caused by cerebral ischemia is demonstrated by the efficacy of antioxidant treatment in attenuating caspase-3 activation, DNA fragmentation and lesion size in mice with cortical infarction (Kim et

> al. 2000). Another possible link between oxidative stress and apoptosis is the reduction in the activity of Apurinic apyrimidinic-endonuclease (APE/Ref 1), a nuclear protein which removes the oxygen radical induced AP site in oxidized DNA and regulates transcriptional factors, such as Ap-1, which are redox sensitive (Bennet et al. 1997). Finally oxidative stress has been shown to activate several transcription factors, particularly NF-kB (Dalton et al. 1999). NF-kb is activated after both transient (Gabriel C. et al. 1999) and permanent ischemia (Seegers et al 2000). NF-kb can then induce the transcription of pro-apoptotic genes (Barkett and Gilmore 1999). (Antonio 2001)

The last disease that I would like to introduce to you briefly in this chapter is a bit different from the others. In fact, most people don't even think of it as a disease, and some even celebrate it. I'd like to discuss with you the biological factors of aging. You may recall from chapter 1 the multiple ways to characterize aging and its biological and environmental factors. To briefly reintroduce these theories, consider the following statement from the National Institutes of Health:

> Some have argued that there is an "aging gene," which manifests itself in one of two ways. First, it is claimed that aging is a genetically programmed means to limit population size and avoid overcrowding. Second, aging is an adaptive process to facilitate the turnover of generations, thus aiding in the adaptation of the species to a changing environment. In contrast, others, such as Thomas Kirkwood (1999), argue that these explanations run counter to one of the basic principles of evolution: Species are programmed to survive, not to die. Kirkwood (1999) and others have proposed a Disposable Soma Theory. This theory assumes that aging is probably caused by the gradual and progressive accumulation of damage in the cells and tissues that comes from the

> need to react and adjust to a changing and demanding environment. Over time, the capacity to react to this changing environment becomes more difficult. With age, physiological systems need more resources and time to adjust to environmental demands. (Santariano 2005)

Aging is a network of biological, genetic, and psychological factors. Gerontologists work to distinguish between normal aging, which may include cosmetic changes or even structural changes in brain volume, and what constitutes disease. For example, a somewhat common stigma attached to old age is a sudden change in disposition that causes irritability, a withdrawn attitude, or depression. "But, an analysis of long-term data from Baltimore Longitudinal Study of Aging (BLSA) showed that an adult's personality generally does not change much after age 30. People who are cheerful and assertive when they are younger will likely be the same when they are 80. The BLSA finding suggests that significant changes in personality are not normal due to aging, but instead may be early signs of disease of dementia" (National Institute on Aging 2008).

Aging starts at the cellular level and affects all organs in the body. Aging differs from person to person, and the effects can range in severity. So what does aging look like internally? Consider damage to DNA. On a daily basis, these structures are subject to millions of events that cause them harm. Luckily, as discussed earlier, our bodies are equipped with antioxidant mechanisms to repair this damage, but with age, some of the damage may remain as the DNA loses the ability to repair itself. Scientists think that this may be a key factor in what causes aging. Throughout life, your body is constantly adapting and going through physiological changes:

> Around the age of 20, lung tissue starts to lose elasticity, and the muscles of the rib cage slowly begin to shrink. As a result, the maximum amount of air you can inhale decreases. In the gut, production of digestive enzymes diminishes, affecting your ability to absorb foods properly

> and maintain a nutritional balance. Blood vessels in your heart accumulate fatty deposits and lose flexibility to varying degrees, resulting in what used to be called the "hardening of the arteries" or atherosclerosis. Over time, women's vaginal fluid production decreases, and sexual tissues atrophy. In men, aging decreases sperm production and the prostate can become enlarged. (National Institute of Aging 2011)

The three main factors that affect how we age are genetics, environment, and behavior. Specifying which genes are responsible for longevity is a difficult task. It is more involved, for example, than identifying which genes are responsible for height and eye color. The pathways and contributing factors associated with longevity genes are still too complex and would take a literal lifetime to measure! In lieu of tackling this obstacle, scientists have taken to studying the longevity genes in simpler animals that have significantly shorter life spans than humans, such as worms and flies. Research has led to an interesting observation: not all the genes that affect longer life are promoters of longevity. Altering or deleting some of these genes actually increases life span. Does this mean some genes function specifically to limit or influence longevity in animals and humans? "The human genetic blueprint or genome consists of approximately 25,000 genes made of approximately 3 billion letters (base pairs) of DNA" (National Institute of Aging 2011). In DNA coding, a minute deviation can be found in about every thousand letters. These polymorphisms are the variations associated with certain traits. People who live for a prolonged period—for example, more than a hundred years—may be more likely to share a specific variant in their code. Similarly, people with another trait or characteristic might share another variant. A direct causal relationship is hard to prove, but the association still offers us some information.

Genes may predispose us to or protect us from certain traits or diseases, but they do not completely determine the story of how we age. Environmental factors and our behavioral patterns also have an important effect on the process of aging and its associated diseases. Common environmental hazards that have a particular effect on aging are extreme temperatures, contaminated water supplies, particle pollution and pesticides, and exposure to lead and mercury. "Medical research reveals that environmental factors play a major role in the majority of cases of Alzheimer's and Parkinson's diseases. Diet, exercise, exposure to toxic chemicals and other environmental pollutants, and socioeconomic stress can alter biochemical pathways influencing the risk of these diseases and other chronic illnesses such as diabetes, obesity, cardiovascular disease, and metabolic syndrome" (National Council on Aging 2012).

Negative environmental factors significantly increase the development of degenerative cognitive diseases like Alzheimer's and dementia compared to genetic predisposition alone. These negative factors include smoking cigarettes, a sedentary lifestyle, excessive alcohol consumption, and a Western diet. The environment can also be classified as a socially constructed variable. Quality of housing, land use, and modes of transportation affect the way we age due to the necessity of adaptation and the risk of injury.

Homeostasis, the internal equilibrium of our bodies, gains support through healthy social bonds and physical activity. Our social networks, families, friends, and romantic relationships affect these things. "A number of studies have indicated that those with stronger social ties experience greater health, functioning, and longevity than those who are socially isolated" (Santariano 2005). The sense of control and efficacy associated with these relationships affects the aging process as well. These relationships promote healthy behaviors through steady access to emotional support, outside knowledge, healthy eating habits, and exercise. "Physical activity is typically classified as either leisure-time physical activity (LTPA) or as utilitarian or everyday activities. Research indicates that men are more physically active than women, and that the overall prevalence of both LTPA and utilitarian walking declines with age. Reasons for the decline seem to vary by gender" (Santariano 2005).

This chapter has been building the foundation of this book. We have explained what free radicals are, how they operate within the body, and the effects they have on bodily organs and their systems. We have also discussed antioxidants and their role in neutralizing free radicals, including the way inadequate supplies of antioxidants can lead to oxidative stress and multiple diseases including aging. We will now introduce the free radical theory of aging, what it means to you, and how it can help you improve your overall health and quality of life.

The Free Radical Theory of Aging

"All available work agrees that, across species, the longer the lifespan, the lower the rate of mitochondrial oxygen radical production. This is true even in animal groups that do not conform to the rate of living theory of aging" (Perez-Campo et al. 1998). Aging as defined by the free radical theory of aging has four main characteristics:

1. It is progressive. The aging process is spread out over the course of an organism's life span.

2. It is endogenous, meaning that it originates within an organism.

3. It is universal. It affects all life-forms.

4. It is deleterious—that is, harmful and damaging. (Beckman and Ames 1998)

Aging happens when cells progressively deteriorate in response to stresses and fail to combat them effectively. This leads to a decrease in an organism's longevity. In reference to the theories of aging mentioned above, the free radical theory of aging (FRTA) is the most noted response to the aging phenomenon. In the mid-1950s, Denham Harman, a biogerontologist and professor, delivered the FRTA in a paper titled "Aging: A Theory Based on Free Radical and Radiation Chemistry" (Harman 1956). It postulated that aging is the result of free radical damage generated from normal metabolic reactions. Initially, he received little recognition from his field for his work due to the common belief that the instability of free radicals left them unable to survive in vivo. However, this changed with the "discovery of the anti-free radical enzyme superoxide dismutase (SOD) by McCord and Fridovich in 1969. If the cell makes something to detoxify free radicals, the free radicals must be produced in vivo. Only four years after the discovery of SOD, Britton Chance's group described the production of hydrogen peroxide in isolated mitochondria. This demonstrated that even during normal mitochondria respiration, oxygen is incompletely reduced and gives rise to highly reactive (and unstable) molecules termed reactive oxygen species (ROS)" (Beckman and Ames 1998).

The use of SOD to locate cellular concentrations of O_2 led Harman to identify mitochondria as the main site for free radical generation and attack, and he adapted his theory, calling it the mitochondrial free radical theory of aging (MFRTA). According to Harman, the buildup of oxidative damage that occurs within the mitochondria is responsible for degenerative diseases and aging. He addresses aging as a progressive side effect of oxidative stress in living things dependent on the metabolic rate of a particular species and its environmental stresses and factors. He drew this conclusion from a comparison between the effects of aging and ionized radiation. "At the time it had recently been discovered that radiolysis of water generates hydroxyl radicals (OH) and early experiments using paramagnetic resonance spectroscopy had identified the presence of OH in living matter" (Beckman and Ames 1998).

Harman used these findings to form the following two hypotheses. Oxygen radicals are generated within the body as an effect of enzymatic redox reactions. The enzymes involved in these reactions—and particularly those that contain iron—directly use molecular oxygen in their processes. In addition, oxidative reactions are propagated and catalyzed by traces of iron and other metals in vivo. Harman further thought that there was a possibility of peroxidation chain reactions occurring in vivo based on the evidence of in vitro polymer chemistry with iron salts enhancing oxidation in organic compounds. Both of these statements have since been proven sound. "A low rate of free radical production can contribute to a low aging rate both in animals that conform to the rate of living (metabolic) theory of aging and in animals with exceptional longevities, like birds and primates. Available research indicates there are at least two main characteristics of longevous species: a high rate of DNA repair together with a low rate of free radical production near DNA" (Perez-Campo et al. 1998).

The debate about the association between antioxidant defense and aging is worth having; both sides provide compelling evidence. In comparative species studies, the role of antioxidants and their capacity levels have been measured to investigate their influence on maximum life span potential (MLSP). "Early work by Cutler tested this association: MLSP correlated positively with SOD but negatively with catalase and GPX. Recently other groups have revisited this hypothesis with similar results. Comparisons between SOD, catalase, GPX, and GSH in the brain, liver and heart from mice, rats, guinea pigs, rabbits, pigs, and cows were carried out and revealed that, (1) SOD and catalase activities correlated positively with MLSP, (2) GSH correlated negatively with MLSP, and (3) GPX actively correlated negatively in the liver and the heart" (Beckman and Ames 1998).

In comparisons with vertebrates of more closely related classes of birds, mammals, fish, and frogs, the correlation between antioxidants and MLSP showed either no relationship or a negative correlation. Another comparative study between two closely related species of mice, the white-footed mouse and the house mouse, showed that "the white footed mouse which lives twice as long had revealed higher levels of SOD, catalase, and GPX in the brain and heart extracts than that of the latter, with the difference in GPX being the most dramatic" (Beckman and Ames 1998). There have also been amazingly promising results with fruit flies treated with curcumin, an antioxidant in the ginger family. "The number of live fruit flies was noted daily and mean lifespan determined for each treatment group. A significant ($P \leq 0.05$) increase in mean lifespan was noted only for the fruit flies maintained on 1.0 mg of curcumin/gram of media; this effect was reversed by addition of disulfiram. These results demonstrate that dietary curcumin prolongs lifespan and that this effect is associated with enhanced superoxide dismutase activity" (Suckow and Suckow 2006).

Ultimately, it is extremely important to mention that the MFRTA is not a statement of causation but of correlation. Scientists are still trying to understand this relationship and how to apply this theory. The free radical theory of aging states that proper supplementation of dietary antioxidants may not allow a person to exceed maximum life span potential, but that person will have a higher chance of living healthily to that maximum life span potential.

In the next chapter, we will break down oxidation and reduction reactions as they relate to free radical production. A direct analysis of the impact these chemical processes have on our cells will allow a complete understanding of how degenerative diseases are formed.

Atoms, molecules, and energy are not bad; we are.

—Alfred Sparman, MD

5

Oxidation-Reduction Reactions

Have you ever wondered why after you slice an apple it starts to turn brown? The answer is oxidation.

The skin of the apple naturally inhibits this reaction, but once the inside flesh is exposed to air, oxidation will occur. Both phenols and polyphenol oxidase, organic chemicals, exist within the apple, but they are usually kept apart. The cutting ruptures this separation and causes a chemical reaction. "The outer skin of an apple acts to protect the inner pulp against injury. But if this layer is damaged, the cells in the pulp release enzymes which react with oxygen to oxidize the damaged cells and form a protective layer against infection. Thus, the brown coloration in an apple occurs in response to damage caused to the apple (similar to blood clotting in humans)" (UCSB Science Line 2015).

The enzyme in fruits and vegetables that causes this brown discoloration, which affects not only their appearance but also their taste and nutritional value, is polyphenol oxidase (PPO). PPO oxidizes the phenolic substrates, and quinones (organic compounds) are produced. They react with each other and form melanin. "Severe browning of plant products arises in stress conditions due to subcellular decompartmentation and oxygen penetration leading to PPO-substrate contact. These reactions are complex given that a large number of monophenolic and/or diphenolic compounds catalyzed by PPO may in turn form a variety of products (quinones and condensation products)" (Yoruk, Ruhiye, and Marshall 2003).

All chemical processes are affected by temperature. If you were to place a sliced apple in the oven, the process would speed up; if you were to place the same apple in the refrigerator, the oxidation process would slow down. Reducing agents are factors that slow down chemical reactions. In this example, you could add sulfates or ascorbic acid, also known as vitamin C, to the exposed parts of the apple. These agents combine with the oxygen and prevent the polyphenol oxidase from reacting.

As mentioned in the previous chapter, oxidation within the human body can have adverse effects once it exceeds a certain threshold, inducing stress on cells and tissues, but it also can have benefits. For example, lipid oxidation is a process in which fatty acids are used to produce energy. The oxidation of fats can help increase metabolism and help with weight management. "The accumulation of fat cells in the body usually leads to an increase in body weight. The fat cells are comprised of fats and excess sugar. Oxidation of these fat cells assists in doing away with the fats and excess sugar. This is a highly beneficial action as it leads to significant weight loss" (Daniel 2011).

Nitric oxide (NO) is a free radical involved in cellular signaling and is most commonly released during exercise. The benefits of NO include a multitude of different cellular activities, but it is most notably recognized for its effects on cardiovascular health. "NO promotes healthy dilation of the veins and arteries so blood can move throughout your body. Plus, it prevents red blood cells from sticking together to create dangerous clots and blockages" (Sinatra 2015).

Despite these useful functions of oxidation, there is reason for caution when it leads to the proliferation of free radicals. It is important to remember that oxidation by itself is not innately or automatically harmful. Generally speaking, ROS and RNS in low quantities promote healthy regulation of cellular signaling and proliferation. However, the buildup of oxidative stress causes degeneration and leads to toxicity. This is why reduction and scavenging of free radicals by intercellular and enzymatic antioxidants is important to keep the body balanced. "Antioxidants can act directly as reducing agents, donating protonic hydrogen to the unpaired oxygen electron or by stabilizing or relocating the free radical electron. In this process the reducing agent becomes oxidized" (Sigma-Aldrich 2007). Because they can become oxidized themselves, it is important to steadily replenish chain-breaking antioxidants like beta-carotene and vitamins C and E.

When these two reactions, oxidation and reduction, work together, the results become vital components of biochemistry and industry. Oxidation-reduction, also known as redox, is the process of a dual electron transfer between two species that exhibit a simultaneous change in oxidation. As the name states, during this reaction, there are two main parties involved, the oxidizing agent and the reducing agent. The oxidizing agent, or oxidant, loses electrons, causing an increase in its own oxidation number. It has been oxidized by the reducer. The reducing agent or reductant, meanwhile, gains electrons, causing its oxidation number to decrease; it has been reduced by the oxidant. Oxidation can happen in three ways: by the addition of oxygen, by the removal of hydrogen, or (most commonly) by the removal of electrons. "Electrons are not stable in the free state, so their removal from a substance (oxidation) must be accompanied by their acceptance by another substance (reduction) hence the reaction is called oxidation-reduction reaction or redox reaction and the involved enzymes are called oxido-reductases" (Aboazma).

Substances with the ability to oxidize are also called electron acceptors; due to their electronegative nature, they tend to hold on to electrons. Reducers, or electron donors, tend to be electropositive and give their electrons away. Chemically, whether an atom or molecule has the ability to oxidize or reduce is dependent on its redox potential. This measurement is taken in voltage and relates to the strength of the relationship between a substance and its electrons. "Electrons are transferred from substances with low redox potential to substances with higher redox potential. This transfer of electrons is an energy yielding process and the amount of energy liberated depends on the redox potential difference between the electron donor and acceptor" (Aboazma). Generally, substances with stronger oxidizing potentials than hydrogen are oxidizers, and those with a lower reducing potential than hydrogen are reducers. A substance's redox potential is somewhat of a rational assumption or theoretical interpretation; metabolic factors like concentration, temperature, pH balance, and pressure can all have an effect on the actual outcome. A common pneumonic device for remembering these redox transactions is OIL RIG, which stands for "Oxygen is lost; reductant is gained."

Reduction

Oxidant + e^- ⟶ Product

(Electrons **gained**; oxidation number **decreases**)

Oxidation

Reductant ⟶ Product + e^-

(Electrons **lost**; oxidation number **increases**)

Two Halves of a Redox Equation (Garnham 2006)

Electrons have a negative charge, so an increase in electrons will lower the net charge; correspondingly, a decrease in electrons will increase the positivity of a charge. Despite what the name may imply, a redox reaction does not necessarily need to involve oxygen; any chemical reaction where there is a change in oxidative state is an oxidative reduction reaction. Think of it as an all-inclusive term. Redox reactions are natural and vital parts of life that take place in a number of processes, such as corrosion, photosynthesis, and respiration. There are five main types of redox reactions. Combination reactions occur when two or more elements join together to form a single product. In a decomposition reaction, the opposite is true: a chemical compound breaks down into its corresponding elements. Single and double displacement, also known as replacement reactions, involve the replacement of one or two elements in the reactants with another element in the product. For example, a generic equation for double displacement would look something like this: AB + CD = AD + CB. Most combustion reactions rely on molecular oxygen. Almost all forms of combustion are exothermic reactions, meaning they produce heat; reactions that produce both heat and light are commonly known as burning. And finally, disproportionate reactions occur when a substance is both reduced and oxidized. An example you may be familiar with is the reaction between hydrogen peroxide and an open wound. The hydrogen peroxide (H_2O_2) decomposes, producing both oxygen (O_2) and water (H_2O). Oxygen is included in every part of this reaction and is thus both oxidized and reduced.

Understanding oxidative states is essential to understanding redox, because oxidized states can tell you what is being reduced and what is being oxidized. The notion derives from the work of a French chemist named Antoine Lavoisier in the late eighteenth century. He recognized that any element could undergo multiple degrees of oxidation. An example of this distinction can be seen in the difference between sulfuric acid (H_2SO_4) and its weaker constituent, sulfurous acid (H_2SO_3). According to Lavoisier and later chemist Jons Berzelius, the oxides within nonmetals were considered to behave as acids, while the oxides in metals functioned as bases. Their reaction formed salts. If a particular element were able to compose a series of oxidants and derive its own salts, these salts would be interchangeable through a network of oxidation and reduction of one or more of their relative oxidants. For example, "Using a modernized version of Berzelius' dualistic formulas, we see that the difference between calcium sulfite [$CaO \cdot SO_2 = CaSO_3$] and calcium sulfate [$CaO \cdot SO_3 = CaSO_4$] was viewed as being literally due to the increased oxidation of the sulfur atom in the acidic oxide component, whereas the difference between ferrous sulfate [$FeO \cdot SO_3 = FeSO_4$] and ferric sulfate [$Fe_2O_3 \cdot 3SO_3 = Fe_2(SO_4)_3$] was instead due to the further oxidation of the iron atom in the basic oxide component" (Jensen 2007).

The connection between oxidation and reduction as it relates to ionic charges was discovered in the field of electrochemistry in the late 1890s and then modernized and completed in the early 1900s. We were able to see that oxidants are those that gain negative charges or release positive charges, and the reverse is true of reductants. Then, in 1915, due to the impact of the ionic bonding model, scientists began to understand that the loss and gain of electrons was responsible for the loss and gain of positive and negative net charges.

"The development of a positive valency by an atom (schematically through the loss of an electron) corresponds to oxidation. When an atom develops a negative valence (schematically through the gain of an electron) it is reduced" (Jensen 2007). This is why even though an atom, molecule, or ion may gain an electron, we still consider it to be reduced—because the negative charge of the electron reduces its oxidative state. The reverse is true with oxidants. They lose an electron or a negative charge, resulting in an increase in their oxidative state. The oxidation state or oxidation number correlates directly with the number of electrons gained or lost during a reaction with another compound. The charge, which can be positive, negative, or neutral, is an inclusive, hypothetical, and convenient tool used to rationalize and visualize the exchange based on the idea that all bonds are ionic (which they aren't—remember the covalent and metallic bonds discussed in chapter 3) and therefore have charge. Redox reactions in covalent bonds are slightly more complex in that you can actually have an oxidation-reduction reaction without any electrons literally being transferred! The critical concept behind redox reactions is a simultaneous change in oxidative state, so in a covalent bond, if the control a substance has over its electrons has been even slightly lost or gained in comparison to its oxidative state before the reaction, it has been oxidized and/or reduced.

So how does this connect with the body? Why are redox reactions important to us? One reason—the production of energy! Consider what it means to transport electrons; the movement of electrons, by definition, is the generation of electricity! Redox reactions create the necessary power for our bodies to undergo aerobic metabolic reactions within the mitochondria of our cells by breaking down glucose and ADP to form ATP. If the body is not able to produce enough ATP, then the muscles and tissues will experience fatigue. Not only that, but ATP is responsible for the following activities:

- biosynthetic reactions
- muscle contraction
- nerve conduction
- active absorption and secretion
- active transport across biological membranes
- activation of monosaccharides, fatty acids, and amino acids
- formation of creatine phosphate, which is the energy stored in muscles
- biosynthesis of cAMP (Aboazma)

During cellular respiration, an organic compound, glucose, reacts with oxygen and ADP to form ATP, water, and carbon dioxide. The ATP (energy) is released and used to keep all the chemical processes going within the body, while the water and carbon dioxide are eliminated through the waste cycle. Hydrogen and carbon bonds within glucose contain a rather large amount of energy; redox reactions carry and transfer the electrons that form these bonds through molecules known as electron carriers. This significant step allows the body to store energy until it is needed for conversion. The two main electron carriers are known as nicotinamide adenine dinucleotide (NAD) and flavin adenine dinucleotide (FAD). They work together to load and unload electrons on the transport chain. NAD is the more important carrier and is the initiator of the electron transport. "The NAD^+ molecule is used to accept electrons (becomes reduced) in several chemical reactions in glycolysis and the Krebs cycle. NAD^+ accepts a hydrogen ion (H^+) and two electrons ($2e^-$), as it becomes reduced to NADH + H^+. The NADH moves to the electron transport chain and donates a pair of electrons (becomes oxidized) to the first compound in the chain. The oxidation of NADH to NAD+ results in the liberation of 53 kcal/mole (under standard conditions)" (Pearson Education Inc.).

One significant difference between NAD and FAD is their ability to hold hydrogen atoms. FAD can accommodate two, while NAD can only handle one at a time. FAD is part of the flavonoid family. It is a protein that has the ability to be either a donor or an acceptor of one or two electrons in a single action; thus, more often than not, it behaves as an intermediary for acceptors and donors. The electron transport chain carries these electrons repeatedly from donor to acceptor through four protein complexes until they reach the final oxygen acceptor. Redox reactions are typically propelled through chains like this, undergoing a series of different reactions. In theory, all redox reactions operate in equilibrium, meaning there are equal amounts of oxidation and reduction between redox couples. However, in biology, that is rarely the case! Due to metabolism, the course of reduction over oxidation is constantly in a state of flux, as it is affected by unsteady levels of concentrations and enzymatic actions. Incomplete reduction reactions generate enzymatic free radicals. Redox cycling is the repeated reduction and oxidation of reactants containing free radicals, more specifically ROS like superoxide. Redox sensitive metals, such as copper and iron, contribute to this production. "Reduction of molecular oxygen is the principal mechanism for ROS formation. The initial product, superoxide, results from the addition of a single electron to molecular oxygen. Superoxide can be rapidly dismutated by superoxide dismutase (SOD), yielding H_2O_2 and O_2, which can be reused to generate superoxide radicals. In the presence of reduced transition metals, H_2O_2 can be converted into the highly reactive hydroxyl radical HO" (Trachootham et al. 2008).

The extent of redox cycling and subsequent oxidative stress at any time depends entirely on the state of the localized cells and tissues, as well as the overall condition of the active redox couples. Redox functions play important roles in both cell survival and cell death. "Increased oxidative stress can either promote cell survival or induce cell death, depending on the cellular context. Genetically unstable cells can adapt to live with the stress by adjusting the level of ROS to the extent that promotes cell survival, leading to the development of cancer. In contrast, normal or aging cells that failed to maintain redox balance are prone to oxidative stress–induced cell death" (Trachootham et al. 2008).

It is necessary to understand that cell survival does not necessarily equal cellular health. The continuous replication of degenerative cells can lead to irregular signaling and defective disorders. Redox cycling reactions affect cell activities and communication in the following ways.

1. Transcriptional Regulation

Redox reactions function as a major system for regulating cell activities. Take proteins, for example. Redox signaling can control protein expressions through interrupting and modifying transcriptions. Transcriptions occur in the first stages of gene expression; it is here that particular portions of DNA are replicated into different forms of RNA. These instances are sensitive to redox due to the amount of cysteine (a type of amino acid) residue located within DNA construction sites. In order for transcription to take place, the DNA structure must uncoil for proper access to the targeted genes. This is called chromatin remodeling. It has been discovered that the enzyme HDAC has the ability to reverse this process and is redox sensitive. So in addition to direct regulation of transcription, high levels of ROS can affect chromatin remodeling.

2. Direct Oxidative Modification

Oxidative modification is a major component of protein-function regulation. These adjustments affect changes in structure and function that can range from slight disturbances to complete fragmentation. Oxidative modification can target multiple types of amino acids with different thresholds; ones containing sulfur seem to be the most likely respondents. The main facilitators of oxidative modification are radicals OH and NO. "The functional outcome of the oxidation depends on the types of modifications and the criticality of the modified amino acid in the protein function. It may lead to either activation or inhibition of the protein activities" (Trachootham et al. 2008).

Protein carbonylation can be produced either through direct oxidative modification or through interaction with the products of oxidation and amino acids. This process usually results in the formation of reactive aldehydes or ketones. Ketones are organic compounds that are produced within the body when there is a shortage of the insulin hormone responsible for aiding in the breakdown of glucose into energy. Individuals with diabetes are at risk for this type of oxidation. When there is not enough insulin, the body will convert fat into energy, resulting in the formation of ketones in your blood. These ketones may then spill into your blood and cause you to become sick. Protein carbonylation is regularly used as an indicator of protein oxidation, because it is easy to detect compared to other forms of oxidation when it accumulates. "Examples of proteins modified by carbonylation, which can impair their functions, include ANT, Hsp, and BCL2" (Trachootham et al. 2008). Adenine nucleotide translocator (ANT) is the most prevalent protein located in the membrane of the mitochondria. It exports and imports ATP and ADP to and from the mitochondrial matrix. Heat shock proteins (HSP) take their name from the fact that it was originally believed that these proteins only activated during exposure to extreme heat. However, we now understand that these cells are produced as a response to extreme stress, whether it is heat, cold, or even wound healing. B-cell lymphoma 2 protein (Bcl-2) regulates apoptosis through either induction or prohibition.

3. Regulation of Redox-Sensitive Interacting Proteins

Various proteins use contact with each other as a way to maintain stability. Protein stability is determined through the thermodynamics of a protein, the rate and way that it folds and unfolds itself. Redox-sensitive interactions may alter proteins' functions negatively through inhibition. Oxidation modification in certain signaling proteins disengages the complex and affects the level and duration of the protein function.

4. Regulation of Redox-Sensitive Modifying Enzymes

Posttranslational modification (PTM) of proteins refers to the process in which, during or after the generation of protein cells, covalent enzymatic processes modify them to produce mature proteins. Phosphorylation, the addition of a phosphate that either activates or deactivates protein activity, is the most common modification that follows PTM. Balance between kinases (enzymes that regulate phosphate transport) and phosphates is what determines the state of the protein. Phosphotyrosine (PTK) is a modification protein that is affected by oxidation.

"Interestingly, whereas thiol oxidation of phosphotyrosine kinase (PTKs) leads to their activation, transient oxidation of protein tyrosine phosphatases (PTPases) inhibits their functions" (Trachootham et al. 2008). PTPases oxidation shifts the balance between kinases and phosphates toward phosphorylation in specified proteins.

5. Regulation of Protein Turnover

As discussed earlier, one component of protein stability is the rate at which they fold and unfold. The folding process is necessary to prepare proteins for engagement in their functions and effective performance. Redox mechanisms have the ability to regulate protein stability and turnover rate. The protein turnover rate is the balance between protein synthesis and protein degradation. When the former is overactive, it puts the body in an anabolic state to form and support lean tissues. When the latter is overactive, it puts the body in a catabolic state to burn lean tissues. In a homeostatic state, protein degradation is usually handled by the proteasome system that works to break down damaged or unnecessary proteins. Under these conditions, the misfolded or damaged proteins are regulated by the protein ubiquitin and the protein complex 26S proteasome. "However, under oxidative stress, although the ubiquitin-activating enzymes (E1) and 26S proteosome are oxidatively inactivated, oxidized proteins may no longer be ubiquitinated and degraded. Instead, such oxidized products can be eradicated by the 20S proteosome in a ubiquitin-independent manner" (Trachootham et al. 2008).

Redox Cycling and Disease

Redox cycling has influence over the mammalian cell cycle. The redox reduction of free radicals at this level of control correlates with the proliferation of disorders and degenerative diseases. The cell cycle goes through three stages: interphase, mitosis, and cytokinesis. These three phases encompass a cell's growth, division, and replication. Studies have shown that the cellular redox environment, the balance between the production of ROS/RNS and their dismutation and removal by antioxidants, goes through fluctuations during different periods of the cell cycle. The results reinforce the hypothesis that there is a bond between metabolic redox oxidation and control over cellular functions. In a study with human adenocarcinoma cells (cancerous tumor cells), researchers observed fluctuations in oxidation levels during mitotic (division) stages and G_1 (growth) stages: "HeLa (human adenocarcinoma) cells synchronized by mitotic shake-off were replated and then harvested at different times after plating for flow-cytometry measurements of the cellular redox environment. The fluorescence of a prooxidant-sensitive dye ($DCFH_2$-DA) was three- to fourfold higher in mitotic cells compared with cells in the G_1 phase. The cellular redox environment increased gradually toward a more-oxidizing environment as G_1 cells moved through the cell cycle. These results suggest that a redox control of the cell cycle regulates progression from one cell-cycle phase to the next" (Sarsour et al. 2009).

There also is evidence that the redox cycle directly influences the cell cycle. This includes the presence of higher levels of antioxidants in the later growth stages and the mitotic stage in comparison to the G_1 stage as well as pharmacologic studies of cell progression that have analyzed the redox environment. "This hypothesis is also supported by a recent report demonstrating significantly higher GSH content in the G_2 and M phases compared with G_1; S-phase cells showed an intermediate redox state. Furthermore, pharmacologic and genetic manipulations of the cellular redox environment perturb normal cell-cycle progression" (Sarsour et al. 2009).

The early embryonic stage of cellular proliferation has a significantly lower oxygen concentration than later stages. This transition from an environment low in oxygen to one that is high in oxygen creates a gradient that affects cellular proliferation both directly and indirectly. What this means is that any defect, such as excess ROS/RNS, within this redox cell cycle can lead to aberrant cell proliferation. This type of proliferation is key to various forms of human pathologies, such as diabetes, cancer, and cardiovascular and neurological diseases. As mentioned earlier, the redox environment is based on the balance between redox products and antioxidants. It has been heavily documented that antioxidants play a role during cell development as well. For example, in a study where MnSOD, an antioxidant that functions to detoxify free radicals produced during mitochondria respiration, was muted in knockout mice, the animal models did not survive past the neonatal stage or were subject to developmental defects. "Homozygous MnSOD-knockout mice survive the embryonic stage of development. However, these mice die after birth of lactic acidemia, cardiomyopathy, and degeneration of the basal ganglia. Developmental defects in MnSOD-knockout mice are associated with damage to mitochondrial aconitase, complex I, and succinate dehydrogenase. In comparison to control mice, defects in MnSOD-knockout mice were very pronounced after oxygen exposure, with a subsequent increase in ROS production" (Sarsour et al. 2009).

In contrast, the common enzyme catalase that performs as an antioxidant as it stimulates the decomposition of hydrogen peroxide (H_2O_2) into oxygen and water was used in a recent study to prolong the life span of mice by 20 percent. The overexpression of catalase within the mitochondria delayed age-related mitochondrial cellular degeneration in those mice. The overexpression of MnSOD and CuZnSOD has also been stated to extend the life span of the *Drosophila* fly. It is clear that the concentrations of redox products and antioxidants strongly affect the cellular redox environment in matters of both preservation and deterioration. Dietary antioxidants as well as their enzymatic relatives play a significant role in free radical management. Consider, for instance, the antioxidant properties of polyphenols. Polyphenols are organic chemicals that foster an abundance of micronutrients in our diet. They can be sourced from a wide variety of fruits and vegetables.

High Polyphenol Foods—Fruits

- apples without skin, apple butter, or applesauce
- apple cider and juice
- apricots
- black or red currants
- blackberries
- blood oranges

- blueberries
- chokeberries
- cranberries
- dates
- elderberries
- gooseberries
- green apples (with skin)
- kiwi
- lemon
- lingonberries
- limes
- mangoes
- marionberries
- nectarines
- oranges, tangelos, tangerines, and so forth (the white pithy stuff is flavonoid-rich)
- peaches
- pears
- plums and prunes
- pomegranates
- quinces
- red or purple grapes
- red apples (with skin)
- raspberries
- rhubarb
- raisins
- strawberries
- sweet or sour cherries

High Polyphenol Foods—Vegetables

- artichokes
- broccoli
- celery (particularly the hearts)
- cherry or grape tomatoes
- corn

- eggplant (aubergine)
- fennel
- garlic
- greens, like kale and turnip
- kohlrabi
- leeks
- lovage
- onions
- parsnips
- raw spinach
- red cabbage
- red and yellow onions
- rutabagas
- scallions
- shallots
- small, spicy peppers
- sweet potatoes
- watercress

High Polyphenol Foods—Legumes, Nuts, and Seeds

- almonds
- cashews
- chickpeas
- dried beans—black beans, red kidney beans, pinto beans, and black-eyed peas
- English peas
- fava beans
- flax seeds
- green peas
- hazelnuts
- lentils
- nut butters
- pecans
- peanuts
- pistachios

- pumpkin seeds
- snap beans
- sunflower seeds
- walnuts (Gene Smart 2015)

The polyphenols in these dietary sources have been linked to the prevention of oxidative stress and related degenerative diseases. Antioxidant supplementation in accordance with a healthy diet also has this effect. "For example, cysteine supplementation in addition to the normal protein diet has shown significant beneficial effects on several parameters relevant to aging, including skeletal muscle functions. *N*-acetylcysteine (NAC), a glutathione precursor, has been shown to protect against oxidative stress–induced neuronal death and thus might delay neurogenerative disease. Melatonin, a physiologic hormone and antioxidant, seems to have a protective effect against neurodegenerative diseases, especially Alzheimer's disease. Phenolic compounds such as resveratrol are potent antioxidants and have been reported to be protective against neuronal apoptosis associated with the pathogenesis of Alzheimer's disease" (Trachootham et al. 2008).

One of the diseases discussed in our last chapter on free radicals was aging. Let's take a look at how redox reactions affect cellular senescence, or the degeneration of cells as they grow and divide.

The free radical theory of aging is on one part of the theory of aging categorized as error theory. An opposing theory, program theory, considers aging to be a preexisting or programmed event in the life cycle. Error theory holds to the belief that has been stated throughout this book that excessive ROS cause critical damage to otherwise healthy cells. "One characteristic of aging is the loss of cellularity and the gradual decline of tissue function. This may be caused by progressive senescence of postmitotic tissues, likely due to chronic damage by ROS. Multiple evidence suggests the involvement of redox imbalance in the aging process" (Sinatra 2015).

As this damage accumulates, it articulates itself through the aging process and results in defects that cause disease. It is also quite possible that redox regulations have some influence over the chronological or sequential process of aging. Quiescent cells (G_0) are the defining characteristics in chronological age. During this state of development, the cells are neither dividing nor preparing for any form of division. Chronological life span potential is dependent on the transit of these cells into the proliferation stage. "In response to mitogenic stimuli, quiescent cells enter the proliferative cycle and subsequently transit back to the quiescent state. This reversible property of cellular quiescence is highly essential to protect the chronological lifespan and avoid aberrant proliferation. MnSOD activity protects the chronological lifespan of normal human skin fibroblasts from age-dependent loss" (Sarsour et al. 2009).

In a study conducted with fibroblasts from human skin, quiescent cells were cultured and incubated for forty to sixty days. The results showed that these cells were no longer able to reenter the proliferation cycle after being replated, or covered. This disallowance was associated with high levels of p16 and a decrease of p21. These proteins are a type of enzyme responsible for cellular progression through the cell cycle. They are called cyclins, and they are related to tumor suppression (p16) and cell division regulation (p21). However, with the overexpression of MnSOD activity, age-related accumulation of p16 declined, and p21 increased to a higher level, restoring the quiescent fibroblasts' ability to enter the proliferation cycle! MnSOD activity also has been found capable of controlling ROS signaling in relation to cyclin D1, which is associated with cellular progression through the G1 stage, and cyclin B1 is a regulatory protein associated with mitosis. "MnSOD activity has been shown to regulate a ROS switch favoring a superoxide signal regulating the proliferative cycle and a hydrogen peroxide signal supporting quiescent growth. Higher levels of MnSOD activity were associated with quiescence, whereas lower levels support proliferation. MnSOD activity–regulates transitions between quiescent and proliferative growth was associated with changes in cyclin D1 and cyclin B1 protein levels. These results support the hypothesis that MnSOD activity could maintain a redox-balance protecting the chronological lifespan" (Sarsour et al. 2009).

To conclude, redox regulations not only affect the health of the cells but also their overall cycle and developmental process. In the next chapter, we will continue our discussion of antioxidants to further understand why they are important in maintaining cellular health and how they defend our cells against oxidation and degenerative disease.

Embrace the positive forces, and the negative will find another host.

—Alfred Sparman, MD

6

Antioxidants

Throughout this book, we have discussed the biochemical capacities of antioxidants and their beneficial properties to cellular systems. Remember that an antioxidant is any substance that has the ability to counteract or inhibit oxidation while maintaining stability. The power is in the action: *antioxidant* specifically denotes the characteristic, not necessarily the substance itself. The various types and functions of antioxidants provide a wide array of proven benefits that are consistently capable of delivering protection to the body from internal and external sources of oxidative stress. The body consumes oxygen as a primary source of energy. A complex system with many different types of antioxidants can counteract the stress caused by free radicals. Humans, understanding the preservative nature of antioxidants, can harness them to provide treatments through functional diets that work to counteract degenerative diseases and disorders caused by oxidation.

Evolution

In the beginning, the evolution of oxygen-producing bacteria, cyanobacteria or cyanophyta, directly shaped the current biodiversity of today's living organisms. Through photosynthesis, cyanobacteria produce oxygen gas as a byproduct. Sandstone fossil deposits containing hematite, the mineral form of iron oxide, place the prokaryotic emersion of cyanobacteria at approximately 2.5 million years ago. We know this because in sample fossils predating these findings the contrasts in mineral deposists are dependent on the presence of molecular oxygen. "Conglomerate rocks occur that contain detrital grains of pyrite and urininite deposited in shallow-water deltaic settings, minerals that in the presence of molecular oxygen are rapidly converted to their oxidized forms, for pyrite (FeS_2) to the mineral hematite (Fe_2O_3) and for urininite (UO_2) to its soluble more-oxidized from UO_{4-}. If there had been appreciable oxygen in the overlaying atmosphere when these shallow sediments were laid down, hematite rather than pyrite would occur in these conglomerates, and urininite would have oxygenized and been dissolved" (Whitton 2000).

The atmospheric influence of molecular oxygen determined which species would survive and which would perish due to their inability to adapt to the changing environment. "It is currently believed that the oxygen concentration in Earth's atmosphere may have remained at 1% of its present level until approximately 2 billion years ago, after which the concentration gradually increased to its present value (about 21%) with the increasing success of photosynthetic life forms" (Venturi and Venturi 2007).

Oxygen levels in the atmosphere were high enough to shape evolution in favor of organisms that were dependent on its use. This contributed to the adaptation of metabolic respiration as a prime source of energy among multicellular organisms. Consequentially, it is hypothesized that oxidative stress among both aerobic and anaerobic life-forms developed naturally alongside these oxygen-producing photosynthetic organisms. "The appearance of oxygen in the atmosphere enabled respiratory metabolism and efficient energy generation systems which use molecular oxygen (O_2) as a final electron acceptor, which led to the formation of reactive oxygen species (ROS) in cells" (Temple et al. 2005). Although, atmospheric oxygen is relatively nonreactive, it can give rise to reactive oxygen intermediates which include superoxide ($O_2^{\bullet-}$), hydrogen peroxide

(H_2O_2), hydroxyl radical ($^{\bullet}OH$), and singlet oxygen ($^1O_2$)" (Peroni et al. 2007).

The redox cycle of iron is a principal component of cellular respiration and oxygen transportation. Iron has the potential to become toxic due to its ability to generate ROS species through its partial reduction of oxygen gas by ferrous iron. ROS and oxidative stress, as we know, work to attack cells and inhibit their functions to the point of cellular death and can ultimately lead to disease; to counteract this problem, evolution provided organisms with antioxidant defenses in both enzymatic and dietary forms. "Protective antioxidant enzyme systems consist primarily of superoxide dismutase (SOD), glutathione peroxidase (GPx), catalase and peroxiredoxins. In addition to these endogenous systems, exogenous dietary antioxidants may help to prevent oxidative stress. In particular, mineral antioxidants of marine origin, present in primitive sea, as some reduced compounds of Rubidium, Vanadium, Zinc, lron, Cuprum, Molybdenum, Selenium, lodine (l), etc. which play an important role in electron transfer and in redox chemical reactions" (Venturi and Venturi 2007).

Notice that iron is included in this list as an antioxidant. This may be confusing, considering that ferrous iron aids in the production of free radicals. To clarify, many of these trace elements are prime participants in redox and mettaloenzymes. A mettaloenzyme is a metal, such as ferrous iron, paired with a protein to create an enzyme that allows the metal to function while restraining its potential for harm. The antioxidant nature of iron depends on its behavior; when it passes electrons to ROS species, producing water, it qualifies as an antioxidant.

In the evolution of antioxidants, the first steps are hypothesized to have been simple physical constraints to protect defenseless and vulnerable components. For example, chlorophyll is an antioxidant produced by plants and used during photosynthesis that combats incomplete redox reactions. The relationship between antioxidants and oxidation is so remarkably balanced due to their simultaneous evolution. The human body demonstrates this relationship; we too have evolved to cooperate with the paradoxical nature of oxygen-based reactions. The early stages of human life depended on a natural diet high in antioxidants. "Plant-based, antioxidant-rich foods traditionally formed the major part of the human diet, and plant-based dietary antioxidants are hypothesized to have an important role in maintaining human health" (Venturi and Venturi 2007).

The increasing popularity of diets that lack fruits, vegetables, and whole grains, relying instead on processed foods high in fat and sugar, means that human beings are more likely to be exposed to oxidation and resulting diseases that cause the degeneration of healthy cells. The role that antioxidants play was discovered as a result of their use in the preservation of lipids from oxidation. In fact, the US Food and Drug Administration (FDA) ignores the complexities of antioxidants and their health benefits and focuses mainly on their preservation properties. The FDA defines antioxidants as "Substances used to preserve food by retarding deterioration, rancidity, or discoloration due to oxidation" (Wanasundara 2015).

Dietary antioxidants may exist naturally in food sources and can be added chemically to retard oxidation. Originally, researchers sourced antioxidants for their experiments naturally. For example, "In 1852, Wright reported that elm bark was effective in preserving butterfat and lard. Chevreul showed that the wood of oak, poplar, and pine (in the order of decreasing efficacy) retarded the drying of linseed oil films applied on them, and on all three, it took a much longer time to dry than on glass" (Wanasundara 2015). Adding man-made chemicals, particularly phenolic compounds, to food lipids to prevent oxidation started the theoretical discussion that led to the current field of antioxidant research.

Solubility

Antioxidants can be fat soluble, water soluble, or both. Fat-soluble antioxidants, or hydrophobic antioxidants, can be stored in the body for prolonged periods of time. Generally speaking, fat-soluble antioxidants' main role is to protect cell membranes from lipid peroxidation. Lipid peroxidation occurs when free radicals take electrons from lipids in the cell membrane and cause cell damage or death. Water-soluble antioxidants, or hydrophilic antioxidants, react with oxidants that occur in the blood plasma or cell cytosol. This is the part of the cell within the membrane that is made up of the cytoplasm and contains cellular organelles.

System Functions

There are multiple ways to classify antioxidants. In this particular section, we will focus on their systems—the how of free radical defense. There are two main types: primary or chain-breaking antioxidants and secondary or preventative antioxidants. Chelating antioxidants can deactivate the prooxidant nature of some metals by interrupting redox cycling. Multiple-function antioxidants, as the name implies, tend to use a combination of methods. These are not to be confused with synergist antioxidants, whose overall activity is enhanced by combination with another antioxidant or compound. There are also enzymatic antioxidants, which the body produces naturally.

Primary antioxidants are an important line of defense against free radical activity. Their chemical makeup allows them to act either as free radical acceptors or as scavengers. They perform best when introduced at the initiation stage. These antioxidants work to completely inhibit or at least delay the propagation or initiation of free radical chain reactions. For example, flavonoids are plant-based, water-soluble antioxidants that fit this description. The scavenging function of these primary antioxidants works through their rapid release of hydrogen atoms. The effectiveness of the antioxidant in these cases depends on electron resonance dissociation and the strength of the hydrogen bond. "In general, the radical-scavenging activity of flavonoids depends on the molecular structure and the substitution pattern of hydroxyl groups (i.e., on the availability of phenolic hydrogens) and on the possibility of stabilization of the resulting phenoxyl radicals via hydrogen bonding or by expanded electron delocalization" (Amic et al. 2003).

Primary antioxidants mostly react with hydroperoxyl radicals and lipids, producing more stable subspecies that are less inclined to partake in free radical propagation due to their low level of reactivity. "The antioxidant radicals are relatively stable so that they do not initiate a chain or free radical propagating autoxidation reaction unless present in very large quantities. These free radical interceptors react with peroxy radicals (ROO•) to stop chain propagation; thus, they inhibit the formation of peroxides. Also, the reaction with alkoxy radicals (RO•) decreases the decomposition of hydroperoxides to harmful degradation products" (Wanasundara 2015).

Secondary antioxidants do not stabilize free radicals. Instead, they work to slow down the rate of oxidation. These are also known as preventative antioxidants and tend to amplify the effectiveness of primary antioxidants. They are capable of performing a number of duties, including the decomposition of radical species, oxygen scavenging, radiation absorption, and—most notably—acting as chelating agents for reactive metals. Chelation is the way particular compounds bind to metal ions. When chelating agents behave as antioxidants, they raise the initial energy needed for free radical reactions. This lowers the redox potential and stabilizes the metal ion. Oxygen scavengers are also an important subclass of antioxidants, because oxygen is necessary in the oxidation process. Carotenoids, the high-pigment molecules in fruits and plants that give them their color, are capable of deactivating oxygen in its excited state. Singlet oxygen (excited oxygen), once deactivated, no longer has the potential to oxidize lipids. "Carotenoids are capable of inactivating photoactivated sensitizers by physically absorbing their energy to form the excited state of the carotenoid. Later, the excited state carotenoid returns to ground state by transferring energy to the surrounding solvent. Other compounds found in food, including amino acids, peptides, proteins, phenolics, urates, and ascorbates also can quench singlet oxygen" (Wanasundara 2015).

Synergism, in the context of antioxidants, refers to the amplified effectiveness of two antioxidants or an antioxidant and a compound when working together as opposed to individually. Multiple primary antioxidants may interact with each other, or primary antioxidants may act in cooperation with metal chelators/peroxy scavengers. Chelator and scavenger combinations work exceptionally well. The chelator inhibits the metal ion oxidation, reducing the production of radicals. This results in creating an antioxidant-potent environment and allows the scavenger potential to increase.

Enzymatic antioxidants, as discussed in the chapter on free radicals and the free radical theory of aging, can be broken down into four main enzymes: superoxide dismutase (SOD), glutathione peroxidase (GSHpx), glutathione reductase, and catalases (CAT). Together, they form the body's natural internal defense against oxidation and cell deterioration. "These enzymes also require co-factors such as selenium, iron, copper, zinc, and manganese for optimum catalytic activity. It has been suggested that an inadequate dietary intake of these trace minerals may compromise the effectiveness of these antioxidant defense mechanisms. The consumption and absorption of these important trace minerals may decrease with aging" (Percival 1998).

The enzymatic reduction pathway's initial reactor is SOD, which converts the (O_2) radical into hydrogen peroxide (H_2O_2). From there, peroxidase and catalase break down the product into two water molecules. SOD is an enzyme that can be found in all living things and behaves like a catalyst for specific chemical reactions. There are certain cases where SOD can be injected into the body as a shot and used to treat medical conditions like inflammation. SOD can also be taken as a dietary supplement and has been speculated to aid in the removal of wrinkles, rebuilding tissue, and the overall extension of life. In one study led by the research group of Xavier Leverve, twenty volunteers received either melon SOD extract or a placebo to test the effectiveness of SOD combined with gliadin in the prevention of cellular damage caused by oxidative stress.

> The experimental design included a daily dose of SOD–gliadin (1000 U SOD activity) or placebo for 14 days before exposure to 100% O_2 in a hyperbaric chamber for 60 min. Hyperbaric oxygen (HBO) therapy is used to treat a variety of diseases, however, it also may cause adverse effects. DNA damage is a well-documented side effect of HBO and can be monitored using a single-cell gel electrophoresis or "comet assay" Therefore, participants in this study were tested for DNA damage, and the results showed a significant decrease in DNA strand breaks in the SOD–gliadin-treated group compared with the placebo group. Additionally, treated participants also demonstrated a diminished concentration of plasma markers for oxidative stress. (Ramono 2015)

Glutathione enzymes are water-soluble antioxidants synthesized by amino acids in the liver. They play various roles throughout the body, including tissue building, chemical and protein production, and assisting the immune system. Glutathiones can be classified into four separate enzyme variations, including glutathione, glutathione peroxidase, glutathione reductase, and glutathione S-transferase. Glutathione peroxidase contains four selenium cofactors that catalyze the breakdown of organic hydroperoxides and H_2O_2. Glutathione S-transferases have a high rate of reactivity with lipid peroxides and are found at high concentrations in the liver. These enzymes are able to annihilate ROS on sight and are core participants in xenobiotic metabolism (Percival 1998). Xenobiotics are foreign chemical compounds within a living organism that the body does not recognize, such as poisons or drugs. Glutathione is called into action when there is an unseemly amount of xenobiotics present.

Glutathione conjugates, or temporarily binds, with the chemical compound, disabling and neutralizing it. Glutathione and vitamin C work well together to neutralize free radicals as well. Health care officials may provide supplementary glutathione as an oral medication or an injection. Studies have shown that glutathione, used as a topical cream, is beneficial for improving and whitening the skin. One study conducted with thirty healthy, adult women yielded the following results: "The skin melanin index was significantly lower with GSSG (oxidised glutathione) treatment than with placebo from the early weeks after the start of the trial through to the end of the study period (at 10 weeks, $P < 0.001$). In addition, in the latter half of the study period GSSG-treated sites had significant increases in moisture content of the stratum corneum, suppression of wrinkle formation, and improvement in skin smoothness. There were no marked adverse effects from GSSG application" (Watanabe et al. 2014).

Catalases are enzymes that catalyze reactions by converting their substrate hydrogen peroxide into water with the help of either iron or manganese. These enzymes are found in almost all organisms that live in environments exposed to oxygen. They defend cells and tissues from damage caused by peroxide and are mostly found in the liver.

Another important antioxidant enzyme is lipoic acid, a sulfurous molecule that decarbonates oxidative alpha-keto acids and helps amplify the effects of other enzymatic enzymes! This enzyme is both lipid- and water-soluble and is able to neutralize ROS in both environments. Often called the universal antioxidant, this super enzyme has been proven to produce significant results in cardiovascular health, diabetes, eye health, neuropathy, bone density, and other health issues brought on or propagated through oxidative stress. The uniqueness of this antioxidant's range of solubility contributes to its potency. "Thanks to these qualities, it is easily absorbed and transported into many organs and systems within the body, for example, the brain, liver, and nerves. Contrast this with antioxidants such as vitamin C, which is not very lipid-soluble (so it is not able to penetrate the lipid wall of cell membranes very well), or vitamin E, which is not very water-soluble. When lipoic acid is combined with these antioxidants, the body's ability to fight free radicals is greatly increased. In fact, lipoic acid helps to regenerate vitamins C and E" (Carr and Frei 2010).

Enzyme Regulation

The initial stage of defense against free radicals and oxidative stress comes from antioxidant enzymes. The regulation of these enzymes is heavily dependent on the oxidative state of the cell and gene expressions. Oxygen deprivation can affect the coordination of antioxidant activities just as overexposure to oxidation can. A balanced environment is key to stabilizing cell maintenance, but the body is prepared to account for these stresses. Oxidative stress affects the way enzymes respond. For example, upon exposure, glutathione's oxidated form, GSSG, maintains redox balance and becomes a primary source of stress signaling. "Functioning of GSH as an antioxidant under oxidative stress has received much attention during the last decade. A central nucleophilic cysteine residue is responsible for the high reductive potential of GSH. It scavenges cytotoxic H_2O_2, and reacts non-enzymatically with other ROS: singlet oxygen, superoxide radical and hydroxyl radical (Larson, 1988). The central role of GSH in the antioxidative defense is due to its ability to regenerate another powerful water-soluble antioxidant, ascorbic acid, via the ascorbate–glutathione cycle" (Blokhina, Virolainen, and Fagerstedt 2003).

A number of things can modulate enzyme activity, and hormones are one of those things. Some behave as antioxidants—melatonin, for example—and increase antioxidant activity. Melatonin is a hormone that primes our sleep cycle and sets the rhythm for other activities, such as blood pressure regulation and reproduction. It can permeate membranes and blood barriers easily, which ups its antioxidant potential. "Melatonin has been shown to markedly protect both membrane lipids and nuclear DNA from oxidative damage. Melatonin can directly neutralize several ROS, including hydrogen peroxide. It can also stimulate various antioxidant enzymes, including catalase, either by increasing their activity or by stimulating gene expression for these enzymes. The decrease in melatonin levels observed with age correlates with an increase in neurogenerative disorders such as Parkinson's disease, Alzheimer's disease, Huntington's disease and stroke, all of which may involve oxidative stress. In general, the production of ROS increases with aging and is associated with DNA damage to the tissues" ("Catalase").

As an added bonus, melatonin does not undergo redox cycling! Some antioxidants can behave as prooxidants when triggered by redox cycling and actually promote free radical production. However, melatonin does not behave in this way. "Melatonin, as an electron-rich molecule, may interact with free radicals via an additive reaction to form several stable end-products which are excreted in the urine. Melatonin does not undergo redox cycling and, thus, does not promote oxidation as shown under a variety of experimental conditions. From this point of view, melatonin can be considered a suicidal or terminal antioxidant which distinguishes it from the opportunistic antioxidants" (Tan et al. 2000).

Some regulatory factors that affect overall enzyme function depend on the organ(s) involved as well as environmental and developmental properties. Any of the following variables can directly impact antioxidant activity and each other:

- the particular organ
- developmental stage
- surrounding cofactors
- temperature
- age

Organ specificity is important to mention, because different tissues, especially in higher-order animals, may exhibit certain idiosyncratic responses or sensitivities that impact regulatory factors of enzymatic antioxidants that are unique to that organ. The lungs, for obvious reasons, are the organs most susceptible to hyperoxia, or overoxidation. "An estimated 5–10% of total O_2 consumed by rat lungs goes to form reduced oxygen species" (Harris 1992). Interestingly, newborn rats resist the negative effects of hyperoxia more effectively than adult rats. Hyperoxia is known to stimulate an enzymatic antioxidant response, especially to SOD, catalase, and GSH-Px, both in cell cultures and in vivo. Researchers found a higher concentration in neonatal rats. The difference in antioxidant concentration appears to be the deciding factor in the level of resistance. Hyperoxia also catalyzes the formation of CuZnSOD, obstructing regular tissue expressions executed by the enzyme. Normally, CuZnSOD levels remain constant throughout development except for certain spontaneous elevations. This reaction parallels a response seen in the MnSOD of bacteria sent into stress by O_2 exposure. "In fact, the apparent parallels in the two systems prompted consideration that O_2 could be the inducing factor that creates the elevated levels of enzymes in hyperoxic lung" (Harris 1992).

The brain is another organ highly susceptible to regulatory enzymatic revisions by oxidants. Normally, there are low to moderate amounts of catalase, SOD, and GSH-Px in the brain. The role of CuZnSOD in brain development is both necessary and harmful. Brain tissue benefits from CuZnSOD, as it protects the organ from certain degenerative diseases. "Increasing the CuZnSOD during development is a necessity for the brain especially when protection of preferentially vulnerable brain neurons to neurodegenerative diseases such as Alzheimer's disease poses a threat at all stages of brain development" (Harris 1992).

However, incremental increases of CuZnSOD can become toxic when stimulated by hydrogen peroxide. "Ceballos-Picot et al. showed that transgenic mice that carried the human CuZnSOD gene and had twice the CuZnSOD activity in brain neurons had significantly increased peroxidation in the pyramidal cells and the gyrus dentate. The experiments support the hypothesis that trisomy 21 cells associated with the abnormalities of Down's syndrome work through elevated CuZnSOD" (Harris 1992).

Throughout the life span of an organism, different developmental peaks require different enzymatic activity in both volume and kind. In the brain and liver of rats, it turns out that SOD enzyme regulation depends on both state and organ. "Of the two SOD enzymes, CuZnSOD is the predominant form in rat brain and liver during development. Whereas brain MnSOD increases steadily throughout life, liver MnSOD plateaus at less than 100 days and remains constant" (Harris 1992).

In human lungs, CuZnSOD detected in all stages from fetus to adulthood shows no significant signs of difference. The activity of this enzyme remains the same, while CuZnSOD mRNA shows a steady increase into adulthood. "This study detected similar CuZnSOD positivity in bronchial epithelial cells after the 17th week of gestation and at term. These results are also consistent with a previous study showing that CuZnSOD is mainly localized to the bronchial epithelium of adult human lung" (Kaarteenaho-Wiik and Kinnula 2004).

MnSOD enzyme expression in normal, healthy adult lungs is weak, and extracellular SOD may increase into adulthood. Catalase's expression, in contrast, is present only during the final stages of lung development. It also is the only antioxidant enzyme able to increase throughout the lungs into adulthood at the same rate as mRNA. "This result also indicates that catalase may play an important role against oxidant stress of the human lung and that a lack of catalase may also predispose the preterm lung to oxidant-related injury" (Kaarteenaho-Wiik and Kinnula 2004).

As stated earlier, antioxidant enzymes perform better when they have access to their cofactors. These trace transition metals work to catalyze reactions. Studies have found that diet plays a key role in the effectiveness of SOD's expression. "Modifications to the diet or the application of specific metal ion chelators limit expression of CuZnSOD in tissues. In human infants and numerous species of animals, including sheep, fowl, and rats, the availability of copper has been shown to be a decisive factor controlling CuZnSOD activity" (Life Expression 2007).

Cofactors also have an effect on each other; when the amount of copper in a rat's diet is decreased, there is not only a decrease in CuZnSOD but also a simultaneous decrease in glutathione levels. The mechanics of cofactor regulation include biochemical reactions among other cofactors that may regulate their function and presence. "The lower selenoperoxidase activity is unexpected but shows that Cu_2 affects expression of the selenium-dependent enzyme, possibly by controlling the absorption or tissue retention of selenium. Within the family of SOD enzymes, lowering CuZnSOD activity leads to a paradoxical increase in MnSOD activity. This response has been viewed as one in which the cell compensates for the loss of one SOD form by increasing the other, thereby attempting to keep the total SOD activity at a near-constant level" (Harris 1992).

Temperature is an environmental feature that catalyzes chemical reactivity. In many cases, warmer temperatures heighten these reactions. In a study on the effects of copper and temperature on antioxidant enzymes in black rockfish, researchers found that simultaneous changes in temperature and toxicity exposure led to increased enzyme activity. "After exposure to two copper concentrations (100 and 200 μg/L), GSH levels and GST activities increased significantly, depending on water temperature ($P < 0.05$) in the liver, gill, and kidney of the black rockfish. GPx and SOD activities decreased significantly with both increasing water temperature and copper treatment in the organs of black rockfish ($P < 0.05$). These changes can be seen as initial responses to temperature stress and as a sustained response to copper exposure" (Min et al. 2014).

In general, there is a positive correlation between temperature and the rate of chemical reactions. As the former increases, the latter does as well; you can loosely estimate that with every ten-degree temperature rise (in Celsius), the reaction rate will double. "Overall, these findings support the notion that higher levels of antioxidant defense are necessary in many ectotherms at warmer body temperatures, possibly in response to increased rates of ROS production and/or increased ROS-induced damage. In fact, water temperature affects many chemical and biological processes, including the amount of dissolved oxygen in water, the rate of chemical reactions, and the mobility and metabolism of organisms as well as their sensitivity to toxic substances, parasites, and disease" (Min et al. 2014).

Aging also has an effect on the patterns of antioxidant regulation. Multiple studies show that aging results in a decrease of MnSOD activity. "For many enzyme molecules, their synthesis may be decreased with age; however, their turnover is also lowered, so that apparent enzyme unit numbers, and thus their activities, can be maintained marginally at young levels and even in old age" (Kitani 2007).

These findings, however, tend to be somewhat controversial, because different researchers have not confined their studies to the same developmental periods and types of cells. Nonetheless, the evidence is still relevant for discussion. In the cases studied, the difference in SOD levels was so staggering that age revealed itself as a noteworthy variable. "For example, with the exception of a significant decrease in liver, early results supported no change in MnSOD activity in most organs of murine animals with aging. However, a later study with select populations of mitochondria isolated from rat brain found significant decreases in the MnSOD activity in light and heavy synaptic mitochondria. More recently, De and Darad examined rats at 3, 12, and 24 months and found a strong decrease in CuZnSOD activity in liver, whereas catalase activity became enhanced with age" (Harris 1992).

In a case study where female mice were separated into groups by age and fed a protein-free diet followed by their normal diet, the livers of the young mice underwent an enzymatic disturbance after returning to the normal diet. The results went on to prove that the enzymatic activities in young mice were more resilient than those in older mice. The glutathione S-transferase (GST) in the young mice livers shot up exponentially and then, in just two days' time, came back down to their usual levels. The older mice, however, did not produce the same results; their GST values returned slowly to their normal levels. "Thus, a linear decline of GST enzyme activities with age was clearly observed only in the recovery phase after a normal diet refeeding, rather than at physiological levels or at bottom levels by PFD feeding (Carrillo et al. 2002). We concluded that young mice could recover their normal enzyme activities more quickly and more efficiently than old mice. We also confirmed this phenomenon in three other animal models—male mice and rats of both sexes (Carrillo et al. 1990, 1991, 1992b) (Kitani 2007).

This seems to suggest that in response to the relationship between aging and metabolic reactions, enzymes may have evolved in a number of ways to deal with these changes and that aging does hold the potential to disrupt or excite activity.

Major Dietary Antioxidants and Supplements

Vitamin A

Also known as beta-carotene, vitamin A can be classified as a retinol in its animal form or a carotenoid, a class of molecular pigments found in fruits, vegetables, and some algae and bacteria. This vitamin supports vision health, especially in dim lighting. It also benefits your immune system and keeps the lining of some body parts, such as your nose, free from infection. Vitamin A is a fat-soluble antioxidant that is metabolized from retinol and carotenoids. "Retinols—the vitamin A found in animal-source foods—require very little work by the body in order to convert it to true vitamin

A. Retinols are sometimes referred to as pre-formed vitamin A or true vitamin A due to the fact that they require such little effort on the part of the body in order for it to be usable" (Nourished Media 2009).

Food Sources

- liver
- fish oils
- milk
- eggs
- brightly colored vegetables and fruits (apricots, carrots, mangoes, spinach, tomatoes, asparagus, okra)

Supplements

Vitamin A is available both in a multivitamin and in a solitary form. It can be prescribed as vitamin A, pure beta-carotene, retinol, or some combination. "About 28%–37% of the general population uses supplements containing vitamin A. Adults who are aged 71 years or older, and children younger than 9, are more likely than members of other age groups to take supplements containing vitamin A" (NIH "Vitamin A" 2013).

The discovery of vitamin A dates back to the early 1800s, when it was noticed that malnourished dogs were susceptible to developing corneal ulcers. Between 1912 and 1917, studies were able to identify dietary aids that were necessary for growth but were not fats, proteins or carbohydrates. In 1918, these "accessory factors" were described as fat soluble and named vitamin A (Semba 2012).

Like most fat-soluble vitamins, vitamin A is essential for healthy cell growth and the maintenance of multiple organs. Vitamin A deficiencies result in side effects that severely affect your vision; pregnant women and children are most susceptible to this deficiency. "The most common symptom of vitamin A deficiency in young children and pregnant women is xerophthalmia. One of the early signs of xerophthalmia is night blindness, or the inability to see in low light or darkness. Vitamin A deficiency is one of the top causes of preventable blindness in children. People with vitamin A deficiency (and, often, xerophthalmia with its characteristic Bitot's spots) tend to have low iron status, which can lead to anemia. Vitamin A deficiency also increases the severity and mortality risk of infections (particularly diarrhea and measles) even before the onset of xerophthalmia" (NIH "Vitamin A" 2013).

Vitamin A and Disease

There are three health disorders in particular in which vitamin A may play a direct role: cancer, measles, and age-specific macular degeneration (AMG).

Because of its participation in cellular division, differentiation, and regulation, studies conducted have examined the association between vitamin A and various cancers. Lung cancer risk, for example, has been shown to be lower in individuals who consume high doses of carotenoids. "Several prospective and retrospective observational studies in current and former smokers, as well as in people who have never smoked, found that higher intakes of carotenoids, fruits and vegetables, or both are associated with a lower risk of lung cancer" (NIH "Vitamin A" 2013).

In prostate cancer, studies have shown that individuals taking beta-carotene supplements exhibit a lower risk of prostate cancer. "Carotene and Retinol Efficacy Trial (CARET) study participants who took daily supplements of beta-carotene and retinyl palmitate had a 35% lower risk of nonaggressive prostate cancer than men not taking the supplements" (NIH "Vitamin A" 2013).

Measles is a fatal disease that has a history of affecting children in low-income or developing countries. Deficiencies in vitamin A can lead to serious cases that may result in loss of vision.

The body needs vitamin A to maintain the corneas and other epithelial surfaces, so the lower concentrations of vitamin A associated with measles, especially in people with protein-calories malnutrition, can lead to blindness (NIH "Vitamin A" 2013). With immediate and consistent treatment of high, orally administered doses of vitamin A, mortality rates drop. "A Cochrane review of eight randomized controlled trials of treatment with vitamin A for children with measles found that 200,000 IU of vitamin A on each of two consecutive days reduced mortality from measles in children younger than 2 and mortality due to pneumonia in children" (NIH "Vitamin A" 2013).

In AMG, oxidative stress is suspected of being a root cause in the substantial deterioration of vision loss in the elderly. "The Age-Related Eye Disease Study (AREDS), a large randomized clinical trial, found that participants at high risk of developing advanced AMD (i.e., those with intermediate AMD or those with advanced AMD in one eye) reduced their risk of developing advanced AMD by 25% by taking a daily supplement containing beta-carotene (15 mg), vitamin E (400 IU dl-alpha-tocopheryl acetate), vitamin C (500 mg), zinc (80 mg), and copper (2 mg) for 5 years compared to participants taking a placebo" (NIH "Vitamin A" 2013).

Vitamin C

We will discuss vitamin C here briefly; it will also get its own chapter with further detail later on. Also referred to as ascorbic acid (AA), vitamin C is a water-soluble antioxidant that aids in the maintenance of cellular health and connective tissues and promotes wound healing. A deficiency in vitamin C can lead to scurvy; this disease was prevalent among sailors in the fifteenth century. Due to their routinely extended voyages, their diets suffered from a lack of fresh food. However, protection from scurvy is not the only selling point for vitamin C. Ascorbic acid is an extremely powerful antioxidant that functions in both enzymatic and nonenzymatic reactions. It is capable of donating electrons, contributing to its antioxidant function when triggered. In addition to its role in cell division and progression, during its liquid phase (AA exists primarily in an aqueous state), the enzyme can directly scavenge for free radicals and reduce them to water. "Vitamin C readily scavenges reactive oxygen and nitrogen species, such as superoxide and hydroperoxyl radicals, aqueous peroxyl radicals, singlet oxygen, ozone, peroxynitrite, nitrogen dioxide, nitroxide radicals, and hypochlorous acid (11), thereby effectively protecting other substrates from oxidative damage" (Carr and Frei 2010).

Food Sources

- citrus fruits
- red and green bell peppers
- kiwi
- broccoli
- strawberries
- cantaloupe
- tomatoes

Supplements

Vitamin C is available both in a multivitamin and by itself as a dietary supplement. "The vitamin C in dietary supplements is usually in the form of ascorbic acid, but some supplements have other forms, such as sodium ascorbate, calcium ascorbate, other mineral ascorbates, and ascorbic acid with bioflavonoids" (NIH "Vitamin C" 2015).

Vitamin C and Disease

Vitamin C has been linked to prevention and treatment of cancer and the common cold. People who consume large amounts of fruits and vegetables containing vitamin C may have a lower risk of contracting breast, colon, and lung cancer. A diet high in fruits and vegetables seems to lower heart disease risk as well. Due to cardiovascular disease's association with high levels of oxidative stress and the abundant antioxidant content in these foods, the correlation seems to be self-evident.

Vitamin E

Vitamin E, also known as alpha-tocopherol, is a fat-soluble antioxidant that exists in eight separate chemical forms. Vitamin E is a highly effective free radical scavenger and an essential component in preventing oxidative stress. Beyond its antioxidant properties, this enzyme plays a critical role in cell signaling and functions. "In addition to its activities as an antioxidant, vitamin E is involved in immune function and, as shown primarily by *in vitro* studies of cells, cell signaling, regulation of gene expression, and other metabolic processes" (NIH "Vitamin E" 2013).

It was first discovered in the 1940s that vitamin E has antioxidant properties and that it was a beneficial component in infant nutrition. "The 1940s and the 1950s marked the beginning of interest in the role of vitamin E in infant nutrition. During this period, investigators examined the intestinal absorption of vitamin E in infants and its use for the prevention of hemolysis, retrolental fibroplasia, intracranial hemorrhage, and pulmonary oxygen toxicity. These studies were the forerunners of more recent studies examining possible benefits of vitamin E therapy in premature infants" (Bell 1987).

Food Sources

- vegetable oils (such as wheat germ, corn, sunflower, and soybean oil)
- nuts
- olives
- green vegetables (such as spinach, kale, turnip greens, and mustard greens)

Supplements

Vitamin E contains several different compounds, so it is important to consider the dosage and form when discussing efficacy; supplements containing alpha-tocopherol as opposed to synthetics are most potent. "Most vitamin E-only supplements provide ≥100 IU of the nutrient. These amounts are substantially higher that the RDAs. The 1999–2000 National Health and Nutrition Examination Survey (NHANES) found that 11.3% of adults took vitamin E supplements containing at least 400 IU. Alpha-tocopherol in dietary supplements and fortified foods is often esterified to prolong its shelf life while protecting its antioxidants properties" (NIH "Vitamin E" 2013).

Deficiencies in vitamin E are rare and usually occur in people who have disorders that do not allow them to absorb fat properly. The symptoms of this deficiency range from an impaired immune system to skeletal malformations. Those whose bodies are unable to process vitamin E in its natural form may receive a water-soluble version.

Vitamin E and Disease

Claims for the benefits of vitamin E against disease are supported by its compelling performance as an antioxidant and anti-inflammatory agent. "The mechanisms by which vitamin E might provide this protection include its function as an antioxidant and its roles in anti-inflammatory processes, inhibition of platelet aggregation and immune enhancement" (NIH "Vitamin E" 2013).

Vitamin E may be associated with the delay or prevention of coronary heart disease, cancer, eye disorders, and cognitive decline, among other diseases. Multiple findings support the claim that it lowers the risk of heart disease. "Several observational studies have associated lower rates of heart disease with higher vitamin E intakes. One study of approximately 90,000 nurses found that the incidence of heart disease was 30% to 40% lower in those with the highest intakes of vitamin E, primarily from supplements" (NIH "Vitamin E" 2013).

Vitamin E also may prevent blood clots that could lead to heart attacks, and taken in its dietary form, it can affect mortality rates. "Among a group of 5,133 Finnish men and women followed for a mean of 14 years, higher vitamin E intakes from food were associated with decreased mortality from CHD" (NIH "Vitamin E" 2013).

Vitamin E's effects on cancer are mixed; in some cases, the antioxidant effects are beneficial. For example, "The American Cancer Society conducted an epidemiological study examining the association between vitamin C and vitamin E supplements and bladder cancer mortality. Of the almost one million adults followed between 1982 and 1998, adults who took supplemental vitamin E for 10 years or longer has a reduced risk of death from bladder cancer" (NIH "Vitamin E" 2013). Another study also found that supplements and a diet high in vitamin E could lower the risk of colon cancer: "One study of women in Iowa provides evidence that higher intakes of vitamin E from foods and supplements could decrease the risk of colon cancer, especially in women <65 years of age" (NIH "Vitamin E" 2013).

However, there is not enough evidence to say that vitamin E definitively prevents cancer. In fact, daily intake of large doses of supplementary vitamin E may increase the risk of some cancers.

Both cataracts and age-related macular degeneration are common reasons for vision loss in the elderly. As oxidative stress may be the cause of these disorders, antioxidative treatment has been postulated as a means of defense against degeneration. "The Age-Related Eye Disease Study (AREDS), a large randomized clinical trial, found that participants at high risk of developing advanced AMD (i.e., those with intermediate AMD or those with advanced AMD in one eye) reduced their risk of developing advanced AMD by 25% by taking a daily supplement containing vitamin E (400 IU dl-alpha-tocopheryl acetate), beta-carotene (15 mg), vitamin C (500 mg), zinc (80 mg), and copper (2 mg) compared to participants taking a placebo over 5 years" (NIH "Vitamin E" 2013).

Vitamin E supplements also have been proven to affect cataract formation positively. "Several observational studies have revealed a potential relationship between vitamin E supplements and the risk of cataract formation. One prospective cohorts study found that lens clarity was superior in participants who took vitamin E supplements and those with higher blood levels of the vitamin" (NIH "Vitamin E" 2013).

The brain is supremely vulnerable to oxidative stress, which may result in degenerative diseases that influence cognitive decline. Alzheimer's is a noted cognitive disease linked to oxidative stress. In one trial, researchers tested vitamin E in conjunction with oxidative inhibitors and against placebos to determine its effectiveness against the disease. "A clinical trial in 341 patients with Alzheimer's disease of moderate severity who were randomly assigned to receive a placebo, vitamin E (2,000 IU/day di-alpha tocopherol), a monamine oxidase inhibitor (selegiline), or vitamin E and selegiline. Over 2 years, treatment with vitamin E and selegiline, separately or together, significantly delayed functional deterioration and the need for institutionalization compared to placebo" (NIH "Vitamin E" 2013).

In general, it has been shown that a diet enriched with vitamin E can cause a reduction in the symptoms of cognitive decline. "Vitamin E consumption from foods or supplements was associated with less cognitive decline over 3 years in a prospective cohort study of elderly, free-living individuals aged 65–102 years" (NIH "Vitamin E" 2013).

Selenium

Selenium is an essential, water-soluble trace element with powerful antioxidant properties. "In recent years, Selenium (Se) research has attracted tremendous interest because of its important role in antioxidant selenoproteins for protection against oxidative stress initiated by excess reactive oxygen species (ROS) and reactive nitrogen species (NOS)" (Tinggi 2008).

The soil-based mineral has two forms, inorganic (selenate and selenite) and organic (selenomethione and selenocysteine selenium). It takes part in hormone metabolism, reproduction, DNA synthesis, and of course protection from oxidative stress. Selenium is mostly found in its organic form in the tissues of animals and humans, with a large amount concentrated in the skeletal muscles.

Food Sources

- seafood
- meat
- grains
- dairy products

Plant-based foods may contain selenium depending on the content in their soil. "The levels of Se in foods can vary widely between geographical regions depending on soil Se levels, and these wide variations in soil Se level are reflected in the wide variations found in the Se status of human populations around the world" (NIH "Selenium" 2013).

Supplements

Selenium is available in multivitamin formulas and on its own in both organic and inorganic form. "The human body absorbs more than 90 percent of selenomethionine but only about 50 percent of selenium from selenite" (NIH "Selenium" 2013). Absorption of selenium depends on its form, and it is mainly concentrated in plasma.

Selenium and Disease

Selenium has had promising effects in the prevention and treatment of heart disease, the prevention and reduction of carcinogen-generated cancers, improved immune system function, and thyroid health. In the case of heart disease, oxidation has been linked to the degeneration of lipids, including the low-density lipids (LDL) that are commonly known as "bad cholesterol." LDL builds up on the inside of blood vessel walls, which leads to blockages or clots that can initiate atherogenesis. "One hypothesis is that the presence of high Se as antioxidant selenoenzymes and selenoproteins may help to reduce the production of oxidised LDL and, therefore, would reduce the incidence of heart diseases" (NIH "Selenium" 2013).

Selenium's antioxidant effects have also been shown to reduce oxidation in patients who have suffered ischemia-reperfusion injury. This injury is the result of tissue damage following the return of blood after there has been a loss of oxygen. "A high-Se diet can considerably reduce the effects of reperfusion, and when Se becomes deficient, it can significantly impair intrinsic myocardial tolerance to ischemic insult" (NIH "Selenium" 2013).

The perceived benefits of selenium proteins for patients with heart disease have led to the development of synthetic proteins that replicate their antioxidant properties. "One of the organo-Se compounds that has been extensively investigated for its biological activity is ebselen [2-phenyl-1,2-benzisoselenzol-3(2H)-one]. Ebselen has been shown to exhibit a weak glutathione peroxidase-like activity in vivo and could be a promising cardioprotective agent for myocardial ischemia-reperfusion (I/R) injury" (NIH "Selenium" 2013).

Selenium's effects on DNA, endocrine and immune system health and repair, and apoptosis have led researchers to suspect it has a role in the prevention of cancer. "Epidemiological studies have suggested an inverse association between selenium status and the risk of colorectal, prostate, lung, bladder, skin, esophageal, and gastric cancers. In a Conchrane review of selenium intake, the highest intake category had a 31% lower cancer risk and 45% lower cancer mortality risk as well as a 33% lower risk of bladder cancer and, in men, 22% lower risk of prostate cancer" (NIH "Selenium" 2013).

There is also a compelling case to be made for selenium's use in reducing morbidity rates in cancer: "The hallmark study of Clark and colleagues reported that people who supplemented their diet with selenized yeast, predominantly in the form of selenomethionine (200 µg/day), had a reduction of nearly 50% in overall cancer morbidity. This study, designed as a randomised, double-blind and placebo controlled trial, also showed low incidence of prostate, lung and colon cancers" (NIH "Selenium" 2013).

The immune system naturally generates ROS as a line of defense against xenophobes and other possible toxins; however, once this generation passes a certain threshold, the ROS move beyond regulation and cause damage to cells. Selenium protects host cells from infections and helps the immune system respond effectively to viral infection. "The involvement of Se in the immune system may be associated with a number of mechanisms, including the increased activity of natural killer (NK) cells, the proliferation of T-lymphocytes, increased production of interferon γ, increased high-affinity interleukin-2 receptors, stimulation of vaccine-induced immunity and increased antibody-producing B-cell numbers" (Tinggi 2008).

A clinical study related to the polio virus found that low levels of selenium weakened the immune system's ability to adapt, whereas participants given selenium supplements showed fewer signs of adverse reactions. "A low Se status in humans has been reported to cause a decreased immune response to poliovirus vaccinations. This study also demonstrated that the subjects supplemented with Se showed less mutations in poliovirus than those who received a placebo" (Tinggi 2008).

The thyroid contains a high concentration of selenium. Akin to iodine, selenium is a necessary part of hormonal and metabolic reactions for this organ. A clinical study in women showed a significant relationship between selenium concentrations and thyroid size. "Epidemiological evidence supporting a relationship between selenium levels and thyroid gland function includes an analysis of data on 1,900 participants in the Supplementation en Vitamines et

Minereaux Antioxydants (SU.VI.MAX) study, indicating an inverse relationship between serum selenium concentartions and thyroid volume, risk of goiter, and risk of thyroid tissue damage in people with mild iodine deficiency" (NIH "Selenium" 2013).

Selenium supplements improve the inflammation process in patients with autoimmune thyroid deficiencies. "It has been hypothesised that the possible mechanism of Se in reducing the effect of thyroid autoimmunity may involve the role of GPx and TrxR as antioxidant defense systems for the removal of the ROS and excess hydrogen peroxide (H_2O_2) produced by thyrocytes during thyroid hormone synthesis" (Tinggi 2008).

Severe cases of selenium deficiency may result in cellular necrosis of thyroid cells.

Zinc

Zinc is an essential, water-soluble trace mineral that catalyzes numerous enzymatic reactions in the body. DNA synthesis, immune response, cell division, fertility, and protein synthesis all require sufficient quantities of zinc. As an antioxidant, zinc provides stabilization of sulfhydryls, protecting them from oxidation, and engages in the reduction of hydrogen peroxide and molecular oxygen to more stable hydroxide. Prolonged deficiency of zinc affects the response of the immune system. "In general, long-term deprivation of zinc renders an organism more susceptible to injury induced by a variety of oxidative stresses" (Powell 2000).

Food Sources

- oysters
- red meat
- poultry
- beans
- seafood
- nuts

"Phytates—which are present in whole-grain breads, cereals, legumes, and other foods—bind zinc and inhibit its absorption. Thus, the bioavailability of zinc from grains and plant foods is lower than that from animal foods, although many grain- and plant-based foods are still good sources of zinc" (Powell 2000).

Supplements

As a supplement, the mineral appears in several forms, including zinc sulfate and gluconate. No differences in absorption quality have been found between variations. "Zinc is used for treatment and prevention of zinc deficiency and its consequences, including stunted growth and acute diarrhea in children, and slow wound healing. It is also used for boosting the immune system, treating the common cold, acne, and recurrent ear infections, and preventing lower respiratory infections. It is also used for malaria and other diseases caused by parasites" (Therapeutic Research Center 2015).

Zinc is an important factor in the overall health of immunity cells. Proper supplementation can be used to correct deficiencies that would otherwise result in damage or degeneration. "The body requires zinc to develp and activate T-lymphocytes. Individuals with low zinc levels have shown reduced lymphocyte proliferation response to mitogens and other adverse alterations in immunity that can be corrected by zinc supplementation" (Powell 2000).

Zinc and Disease

Zinc has been found to be beneficial in the treatment of the common cold, reducing the severity and the length of the virus. "Researchers have hypothesized that zinc could reduce the severity and duration of cold symptoms by directly inhibiting rhinovirus binding and replication in the nasal mucosa and suppressing inflammation" (Powell 2000).

There have been some conflicting results pertaining to its effectiveness; however, when administered in lozenge or syrup form, zinc makes direct contact with the rhinovirus in certain areas. "In a randomized, double-blind, placebo-controlled clinical trial, 50 subjects (within 24 hours of developing the common cold) took a zinc acetate lozenge (13.3 mg zinc) or placebo every 2–3 wakeful hours. Compared with placebo, the zinc lonzenges significantly reduced the duration of cold syptoms (cough, nasal discharge and muscle aches)" (Powell 2000).

Similar trials have replicated these results, fortifying previous hypotheses. "More recently, a Cochrane review concluded that 'zinc (lozenges or syrup) is beneficial in reducing the duration and severity of the common cold in healthy people, when taken within 24 hours of onset of symptoms'" (Powell 2000).

Like vitamin E, zinc aids in delaying the development of AMD but does not prevent it. "In a population-based cohort study in the Netherlands, high dietary intake of zinc as well as beta carotene, vitamin C, and vitamin E was associated with reduced risk of AMD in elderly subjects. However, the authors of a systematic review and meta-analysis published in 2007 concluded that zinc is not effective for the primary prevention of early AMD, although zinc might reduce the risk of progression to advanced AMD" (Chong et al. 2007).

Another study also showed that in patients with macular degeneration, zinc supplements helped limit vision loss. "Two other small clinical trials evaluated the effects of supplementation with 200 mg zinc sulfate (providing 45 mg zinc) for 2 years in subjects with drusen or macular degeneration. Zinc supplementation significantly reduced visual acuity loss in one of the studies" (Stur et al. 1996).

Based on this chapter, it should be easy to see that antioxidants can play a powerful role in preventative medicine. The following chapters will direct your focus to three antioxidants whose components are mutually beneficial. Everything stated beforehand has been building up to this point—the introduction of the life pill.

When you are under scrutiny, ask yourself three questions:

1. *Am I doing good?*
2. *Do I believe in what I am doing?*
3. *Am I helping others?*

If all three answers are yes, then success will come.

—Alfred Sparman, MD

7

Moringa Oleifera

The history of *Moringa oleifera* dates back to 150 BC, and ancient kings and queens used the leaves and fruit in their diet to maintain mental alertness and healthy skin. The ancient Mauryan warriors of India drank an elixir made from *Moringa,* as it was believed that the drink would give them extra energy and relieve them of the stress and pain incurred during war (Mahmood, Mugal, and Ul Haq 2010). *Moringa* became a household name for many of us a few years ago, and should you type the name into Google, you will now find more than two hundred thousand hits, including commercial sites selling nutritional products and research supporting folklore practices. It is now sold as a tea, a powder, a drink, and a capsule. It has been claimed that *Moringa* contains seven times more vitamin C than oranges, four times more vitamin A than carrots, four times more calcium than milk, three times more potassium than bananas, and two times more protein than in yogurt. *Moringa oleifera* has been around for centuries and has had numerous names across Africa, Asia, Latin America, and the Caribbean. However, in recent years, we have seen a rebirth of interest in this wonder plant, as many people are on the hunt for an all-natural healer that will assist with both preventative measures and treating ailments. There have been many studies since this rebirth, and they have supported most of the speculation regarding the curative powers of this plant—it may be beneficial for everything from lowering cholesterol and glucose levels to reducing cancerous tumor growth.

Analysis of the phytochemicals in the leaves shows they are a rich source of potassium; calcium; phosphorus; iron; vitamins A and D; essential amino acids; and other antioxidants, including beta-carotene, vitamin C, and flavonoids (Mbikay 2012). The beauty of *Moringa* is its ability to thrive in regions with hot, arid climates with minimal need for watering and other maintenance. Thus, there are programs throughout Africa and in some Asian regions to grow this superfood in backyards.

The Plant

Moringa is native to the Himalayas in northwest India and was planted in many regions due to its ability to flourish in even the driest of conditions. The plant is surprisingly suited to hot and dry regions and remains green year-round. *Moringa* is part of the Moringaceae family, which is made up of approximately thirteen species of *Moringa* trees. However, *Moringa oleifera* is the most widely known. This plant has many names in different parts of the world. In some places, it is called "mother's best friend." In the Nile Valley, it is referred to as *shagara al rauwaq* or "tree for purifying," while in Pakistan it is called *sohanjna.* In some regions, *M. oleifera* has names based on its characteristics or physical appearance. For example, it is called the drumstick tree because of its long, slender seed pods; the horseradish tree because of the taste of its roots; and the ben oil or benzoil tree because of the oil the seeds produce.

The plant grows extremely quickly and can reach a height of fifteen feet within one year, eventually reaching a final height of twenty to twenty-five feet. It is recommended that the tree be pruned regularly to create fuller foliage, because left unpruned, it has a tendency to grow tall and lanky, causing a reduction in its yield and making it difficult to harvest. The tree can be propagated from a seed or a cutting, and the ideal soil is dry and sandy; however, it can tolerate poor soil conditions, including those in coastal areas. *Moringa* trees thrive in the sun and heat, and can be grown as an annual or perennial plant. The *Moringa* tree can also tolerate minimally cold temperatures. The tree is harvested mainly for the leaves. However, the flowers, seeds, and pods all have some nutritional value. It currently grows throughout much of Africa and in many parts of Asia, such as India, Bangladesh, the Philippines, Taiwan, Yemen, and Malaysia. It can also be found in the Caribbean and Latin America; in Barbados, the Dominican Republic, Haiti, the Cayman Islands, Colombia, Trinidad, Puerto Rico, and Suriname. It is also found on some of the islands of Oceana, such as Fiji, Guam, and Palau.

Description

The leaves are feathery, with a pale green color. They are compound tripinnate and average 30 to 60 cm in length, with many small leaflets that are 1 to 2 cm long and 0.3 to 0.6 cm wide.

The Leaves

The flowers are fragrant and bisexual, surrounded by five unequal, thinly veined petals that are either white or creamy white. The flowers are approximately 1 to 1.5 cm long and 2 cm broad. They grow spread out on slender, hairy stalks, later drooping to form the pods or drumstick. Flowering usually occurs within the first six months of planting. The flowers are said to have a similar taste to mushrooms.

The Flowers

The pods are sometimes called drumsticks and are said to taste like asparagus. The hanging fruit is a three-sided brown capsule with an approximate length of 20 to 45 cm and holds dark-brown globular seeds with a diameter of approximately 1 cm. The seeds can be taken from the pod while still green and eaten either boiled or fried. The seeds have three whitish, papery wings. There are no pods in the plant's first year, and for the first two to three years, the yield is generally low.

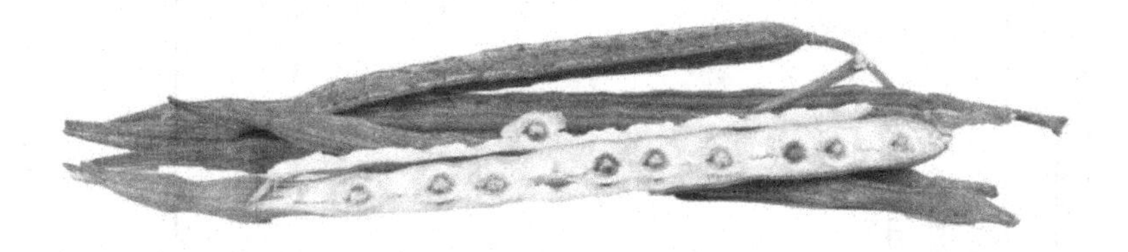

The Pods

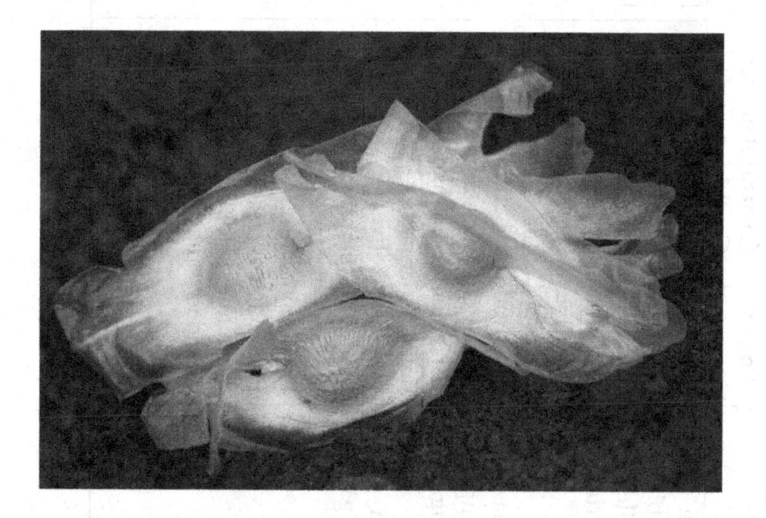

The Seeds

Moringa's Many Names

As mentioned before, *Moringa's* names vary depending on your location.

Africa	Asia	South and Central America, the Caribbean	Oceania
Benin: patima, ewé ilé Burkina Faso: argentiga Cameroon: paizlava, djihiré Chad: kag n'dongue Ethiopia: aleko, haleko Ghana: yevu-ti, zingerindende Kenya: mronge Malawi: cham'mwanba Mali: névrédé Niger: zôgla gandi Nigeria: ewé ilé bagaruwar maka Senegal: neverday, sap-sap Somalia: dangap Sudan: ruwag Tanzania: mlonge Togo: baganlua, yovovoti Zimbabwe: mupulanga	Bangladesh: sajina Burma: dandalonbin Cambodia: ben ailé India: sahjan, murunga, moonga Indonesia: kalor Pakistan: suhanjna Philippines: mulangai Sri Lanka: murunga Taiwan: la mu Thailand: marum Vietnam: chùm ngây	Brazil: cedro Colombia: angela Costa Rica: marango Cuba: palo jeringa Dominican Republic: palo de aceiti El Salvador: teberinto French Guiana: saijhan Guadeloupe: moloko Guatemala: perlas Haiti: benzolive Honduras: maranga calalu Nicaragua: marango Panama: jacinto Puerto Rico: resada Suriname: kelor Trinidad: saijan	Fiji: sajina Guam: katdes Palau: malungkai

(Trees for Life)

History: Claims of Historical Uses

Country	Uses
Guatemala	Skin infections, sores
India	Anemia, anxiety, asthma, blackheads, blood impurities, bronchitis, catarrh, chest congestion, cholera, conjunctivitis, cough, diarrhea, eye and ear infections, fever, glandular swelling, headaches, abnormal blood pressure, hysteria, pain in joints, pimples, psoriasis, respiratory disorders, scurvy, semen deficiency, sore throat, sprain, tuberculosis
Malaysia	Intestinal worms
Nicaragua	Headaches, skin infections, sores
Philippines	Anemia, glandular swelling, lactation
Puerto Rico	Intestinal worms
Senegal	Diabetes, pregnancy, skin infections, sores
Venezuela	Intestinal worms
Other Countries	Colitis, diarrhea, dropsy, dysentery, gonorrhea, jaundice, malaria, stomach ulcers, tumors, urinary disorders, wounds

(Trees for Life)

Global Stories of *Moringa*

Below are some folk tales about and current ayurvedic uses of *Moringa* from around the world (adapted from www.treesforlife.com).

In Guatemala, local doctors crush the leaves into a paste and apply it to sores and skin infections for its rumored healing properties.

In Puerto Rico, an infusion of the flowers is used as an eyewash to cleanse an infection in the eye.

In Nicaragua, believers in home remedies take the buds of the tree and rub them on the temples of someone suffering from a headache, creating a soothing effect.

In Aruba, the seeds are crushed until they form a paste and then spread over warts and other skin afflictions.

Across West Africa, the plant's edible leaves and oils feed livestock; in some regions, doctors prescribe it as a treatment for diabetes, although the effectiveness of this treatment is not scientifically proven.

In Senegal, some doctors still prescribe *Moringa* powder for persons suffering dizziness or weakness; the powder is mixed in with their food to speed recovery.

In ancient Rome, Greece, and Egypt, the oil from *Moringa* seeds was turned into a perfume; it was also believed that spreading the oil across the skin would help to protect the skin from infections and damage.

In India, *Moringa* is popular and highly valued as a nutritious tea; it is also given as a treatment for persons suffering with high blood pressure or anxiety because of its calming effects.

In the Philippines, the leaves are used to make a nutritious soup as a cheap and healthy meal option.

Common Medicinal Uses of Different Parts of *Moringa*

Plant Part	Medicinal Uses
Root	*Moringa* root is an antilithic (it helps with kidney stones and similar ailments), a rubefacient (it helps with pain by dilating arteries and increasing blood flow), a vesicant, and a carminative (it prevents gas or helps with its expulsion of gas). It also is an antifertility measure, an anti-inflammatory, a stimulant in paralytic afflictions. It acts as a cardiac and circulatory tonic, a laxative, an abortifacient, and a rheumatism treatment. It helps with inflammations, articular pains, lower back and kidney pain, and constipation.
Leaf	The leaves are used as a purgative (a laxative), applied as a paste to sores, and rubbed on the temples for headaches. They also are used for piles (hemorrhoids), fevers, sore throat, bronchitis, eye and ear infections, scurvy, and catarrh (inflammation of mucous in the airways or body cavities). *Moringa* leaves also help to control glucose levels and reduce glandular swelling.
Bark	A rubefacient and vesicant, *Moringa's* bark also is used to cure eye disease and for the treatment of delirious patients. It prevents enlargement of the spleen and the formation of tuberculous glands of the neck, destroys tumors, heals ulcers, and provides relief from earaches and toothaches. It also counteracts tubercular activity.
Gum	The plant's gum is used for dental caries (tooth decay) and as an astringent and rubefacient. It provides relief from headaches, fevers, intestinal complaints, dysentery, and asthma and sometimes is used as an abortifacient and to treat syphilis and rheumatism.

Flower	The plant's flower has high medicinal value as a stimulant, aphrodisiac, abortifacient, and cholagogue. It is used to treat inflammations, muscle disease, hysteria, tumors, and enlargement of the spleen. It also lowers serum cholesterol, phospholipids, triglycerides, VLDL (very low-density lipoproteins), and LDL-cholesterol-to-phospholipid ratio. It has decreased lipid profile of the liver, heart, and aorta in hypercholesterolemic rabbits and increased the excretion of fecal cholesterol.
Seed	Seed extract exerts its protective effect by decreasing liver lipid peroxides. Antihypertensive compounds thiocarbamate and isothiocyanate glycosides have been isolated from the acetate phase of the ethanolic extract of *Moringa* pods.

(Anwar et al. 2007)

Nutrient Content and Chemical Breakdown

The contents of *Moringa* plants vary greatly depending on the climate, variety, season, weather conditions, and when they are harvested (for example, a young tree has different nutritients from an older one).

Amino Acid Content of *Moringa* Leaves

All values are per 100 grams of edible portion.

Nutrient	Fresh Leaves	Dried Leaves
Arginine	406.6 mg	1,325 mg
Histidine	149.8 mg	613 mg
Isoleucine	299.6 mg	825 mg
Leucine	492.2 mg	1,950 mg

Lysine	342.4 mg	1,325 mg
Methionine	117.7 mg	350 mg
Phenylalanine	310.3 mg	1,388 mg
Threonine	117.7 mg	1,188 mg
Tryptophan	107 mg	425 mg
Valine	374.5 mg	1,063 mg

Other Nutrients in *Moringa* Leaves

All values are per 100 grams of edible portion.

Nutrient	Fresh Leaves	Dried Leaves
Carotene (Vitamin A)	6.78 mg	18.9 mg
Thiamin (B1)	0.06 mg	2.64 mg
Riboflavin (B2)	0.05 mg	20.5 mg
Niacin (B3)	0.8 mg	8.2 mg
Vitamin C	220 mg	17.3 mg
Calcium	440 mg	2,003 mg
Calories	92 cal	205 cal
Carbohydrates	12.5 g	38.2 g
Copper	0.07 mg	0.57 mg
Fat	1.70 g	2.3 g
Fiber	0.90 g	19.2 g
Iron	0.85 mg	28.2 mg
Magnesium	42 mg	368 mg
Phosphorus	70 mg	204 mg
Potassium	259 mg	1,324 mg
Protein	6.70 g	27.1 g
Zinc	0.16 mg	3.29 mg

Structural motifs and backbones of major phytochemicals found in *M. oleifera* leaves, including glucosinolates and their metabolites, flavonol major derivatives, and phenolic acids. Their known derivatives or metabolites are indicated.

Some bioactive phytochemicals are found in *M. oleifera* leaves. Their structures were obtained from the online PubChem database at www.pubmed.ncbi.nlm.nih.org/.

Current Uses: General and Medicinal

We have reviewed the many historical uses and various folk practices related to *Moringa* and have examined its composition. Now let us review its current uses as well as the supporting evidence, which in many instances serves to reinforce previous theories about *Moringa*'s potential. Over the past thirty years, there has been a significant amount of research conducted on both human and nonhuman test subjects. The majority of the results show an amazing amount of hope for this widely available species.

Diabetes

Diabetes mellitus is one of the most common noncommunicable diseases in the world; according to the *IDF Diabetes Atlas,* as of 2014, an estimated 387 million people were living with the preventable disease. Projections show that by 2035, there will be an additional 205 million cases. In addition, we are seeing an increased number of children diagnosed with type 2 diabetes. Diabetes results from the body's inability to use insulin properly, resulting in an increased level of glucose within the bloodstream. Complications from diabetes are numerous and are often life-threatening. Foot complications often result from poor blood flow. Diabetic ketoacidosis (DKA) is another possible complication. When your cells don't get the glucose they need for energy, your body begins to burn fat for energy, which produces ketones. Ketones are acids that build up in the blood and appear in the urine when your body does not have enough insulin. They are a warning sign that your diabetes is out of control or that you are getting sick. Other possible complications include kidney disease and a higher risk for glaucoma, cataracts, and stroke.

Antioxidants, such as carotenoids, vitamins C and E, and flavonoids play an important role in the reduction of blood glucose in patients by improving the impaired glucose metabolism and decreasing insulin resistance. In a study by Giridhari, Malathi, and Geetha (2011), one hundred type 2 diabetics were selected to observe the antidiabetic properties of drumstick *(Moringa oleifera)* tablets. The study tested glycated hemoglobin levels, that is, HbA1c/HBA1 levels, in the control and experimental groups before and after administering the leaf tablets for fifteen days. HbA1c is a key marker, as it indicates overall blood glucose levels for a period of two to three months. Giridhari, Malathi, and Geetha noted significant decreases in both HbA1c and postprandial blood glucose levels in the experimental group. Many other similar studies similar confirm the viability of *Moringa* in the management of diabetes.

Asthma

Ayurvedic medicine has long used *Moringa* as one of the herbal treatments for asthma. Asthma is one of those chronic illnesses that is typically treated as symptoms arise, and there are currently no pharmaceutical drugs that provide a cure for this worldwide illness. Folklore has long made mention of *Moringa* as a treatment for upper-respiratory-tract infections and other respiratory illnesses. Agrawal and Mehta (2008) conducted a study on human subjects to examine the effects of *Moringa* on asthma. The results suggested that treatment with the *Moringa* for three weeks produced an appreciable decrease in the severity of asthma symptoms and a simultaneous improvement in respiratory functions with no known side effects.

Antifungal, Antibacterial, and Antimicrobial Uses

Numerous researchers have investigated the truth behind the claims in such countries as Guatemala and Senegal that *Moringa* is an effective treatment for skin infections and sores because of its antifungal and antibacterial properties. *Moringa* roots are reported to be rich in antimicrobial agents. They are said to contain an active antibiotic, pterygospermin, which has powerful antibacterial and fungicidal effects. A similar compound has been discovered to be responsible for the antibacterial and fungicidal effects of *Moringa* flowers. The bark extract has also been shown to possess antifungal properties, while the juice from the stem bark has antibacterial effects against staphylococcus aureus. The fresh juice from *Moringa* leaves has been found to inhibit the growth of microorganisms that are pathogenic to human beings (Anwar et al. 2007). Doughari, Pukuma, and De (2007) state that the activities of the plant extracts are comparable to those of pharmaceutical-grade antibiotics, such as ciprofloxacin, cotrimoxazole, and chloramphenicol. This indicates that once purified, the extract could serve as a very potent antibacterial.

Parasites and Detoxification

In Malaysia, *Moringa* is used to cleanse the body of intestinal worms. In a study conducted in 2009, Rastogi et al. compared *Moringa* leaves to the roots of the *Vitex negundo*, more commonly known as the five-leaf chaste tree. The study used powdered extract of the two plants, and round earthworms were the subjects, since they have a similar anatomical and physiological structure to intestinal round worm parasites in humans. *Moringa* exhibited strong anthelmintic properties; that is, it had the ability to cause paralysis in the worms. Paralysis occurred within six and fifteen minutes of administering the powder, while death was observed approximately one hour after administration, strongly supporting the detoxifying benefits of *Moringa*. However, further research is needed before full support can be given to *Moringa* as an alternative worming agent.

Cancer and Tumor Prevention

Cancer is one of the leading causes of death and insurmountable debt globally. Patients spend millions in an effort to slow or reverse tumor growth. Radiation and chemotherapy lead the charge as effective treatments. However, deleterious side effects include changes in the patient's sense of taste, loss of appetite, nausea and vomiting, loss of hair, mouth sores, shortness of breath, and infertility. These are just a few of the side effects, as they may vary depending on the area being treated. They can affect the patient for a short or long period of time. For many decades, researchers have searched for natural or synthetic compounds to delay, inhibit, or reverse the development of cancer. For this reason the spotlight is on plant products, as they are a rich source of phytochemicals. *Moringa* is one of the plants that shows tremendous potential to diminish tumor growth with minimal side effects. Numerous research studies have demonstrated that *Moringa* is a potential source of antitumor activity due to the presence of four phytochemicals. One in particular, niazimicin, shows great potential as a potent antitumor promoter in chemical carcinogenesis (Guevara et al. 1999). Jung (2014) conducted a study using a soluble extract from *Moringa* leaves, confirming that the extract induced apoptosis (cell death), inhibited tumor-cell growth, and lowered the internal level of ROS in several types of human cancer cells. The *Moringa* extract significantly reduced cancer-cell proliferation and invasion. Further studies isolating the phytochemicals need to be done to understand the inhibitory mechanisms on tumor promotion.

Circulatory and Endocrine Disorders: Antihypertensive, Cardiac Tonic, and Thyroid Regulation

Copious studies have supported the benefits of *Moringa* for improving vascular circulation, decreasing blood pressure, and improving cholesterol levels. The majority of the plant is used extensively for treating inflammation and cardiovascular disease. The leaves serve as a rich source of antioxidants, such as beta-carotene, vitamins C and E, and polyphenolics. In Thailand, the young pods, fruits, and leaves have been consumed for more than one hundred years. In addition, the dried root steeped in hot water is taken orally as a cardiotonic, a stimulant against fainting (Chumark et al. 2008). According to studies, the leaf has been known to have a stabilizing effect on blood pressure, possibly owing to the nitrile, mustard oil glycosides, and thiocarbamate glycosides compositions that have been isolated from the leaves (Anwar et al. 2007).

The plant also is a diuretic, and it is believed that this is likely to play a complementary role in its ability to lower blood pressure. Diuretics work by forcing the kidneys to remove more sodium and water from the body, causing the blood vessel walls to relax. Along with its positive impact on blood pressure, *Moringa* has been identified as an agent that helps to reduce LDL and VLDL while increasing much-needed high-density lipoproteins (HDL). Ghasi et al. (2000) examined the effects of the crude extract of *Moringa* leaf on Wistar rats on a high-fat diet; the results conclusively showed a significant reduction in the cholesterol levels of those served the *Moringa* in combination with the high-fat diet compared to those served the high-fat diet alone.

Your thyroid and the hormones it produces play an important role in regulating your body's metabolism, growth and development, and body temperature. During infancy and childhood, adequate production of thyroid hormones is essential for brain development. The two main hormones produced are thyroxine (T4) and triiodothyronine (T3); the hormonal output is then regulated by the thyroid-stimulating hormone (TSH). Normally, these remain balanced; however, habits like smoking and other factors—including age, gender, ethnicity, and even geographical location—can cause a disruption in the thyroid, resulting in either hypothyroidism or hyperthyroidism. Hypothyroidism is when the production of the hormone is low, while hyperthyroidism is as a result of an overproduction of the hormone. Hypothyroidism has been directly linked with hypercholesterolemia—that is, the low production of thyroid hormones, which classically results in high levels of total cholesterol and LDL. Studies have shown that *Moringa* works mostly to regulate the function of the thyroid in instances where persons are suffering with hypothyroidism by increasing T3 and T4 levels while decreasing TSH (Tabassum et al. 2013). Thus far, studies have not been able to pinpoint the mechanisms that cause *Moringa*'s positive action on the thyroid.

Digestive Disorders: Ulcer, Gastritis, and Liver Protection

Ulcers are a relatively common gastrointestinal occurrence globally and can cause severe pain and discomfort. They are triggered when there is a disturbance in the normal balance, either as a product of enhanced aggression or diminished mucosal resistance. Regular use of drugs, irregular food habits, and stress, among other things, can initiate ulcers. Ulcers that are located in the digestive tract—the stomach or duodenum—are referred to as peptic ulcers. These are formed by the presence of acid and peptic activity in the gastric juice along with a breakdown in mucosal defenses. The two most common types of peptic ulcers are gastric ulcers and duodenal ulcers; the names refer to the sites of the ulceration. The main etiological factors associated with peptic ulcers include *Helicobacter pylori*, NSAIDS (nonsteroidal anti-inflammatory drugs), emotional stress, alcohol abuse, and smoking. In folk medicine, *Moringa* leaf tea was used to treat gastric ulcers by the Kani tribe of the Pechiparai Hills in Tamil Nadu, India. The buds from the flowers are widely consumed in Pakistan for their antiulcer properties as well (Vimala and Shoba 2014). Studies have confirmed that the antibiotic and antibacterial properties of the *Moringa* exert an inhibitory effect on various pathogens, such as *Helicobacter pylori* and coliform bacteria, which can trigger diarrhea and ulcer formation. The roots and trees of the *Moringa* have been purported to possess antispasmodic properties, which are attributed to the presence of 4-[α-(L-rhamnosyloxy) benzyl]-o-methyl thiocarbamate. This component forms the traditional basis for *Moringa*'s use for diarrhea.

Hepatoprotection is simply the ability to prevent damage to the liver, which can arise as a result of a variety of causes, including but not limited to hepatitis B and C, long-term alcohol consumption, cirrhosis, malnutrition, and reaction to certain prescriptions and herbal medications. Research has shown that the methanol in

Moringa leaf extract caused antiulcerogenic and hepatoprotective effects in rats; the aqueous leaf extract also showed antiulcer effects (Anwar et al. 2007). Additionally, spasmolytic activity exhibited by different constituents provides the pharmacological basis for the traditional uses of *Moringa* in most gastrointestinal motility disorders. Treatment with *Moringa* is believed to be comparable to treatment with commercially available antacids.

Inflammation: Rheumatism, Joint Pain, Edema, and Arthritis

Joint pain, arthritis, rheumatism, and edema are some of the conditions that occur as a consequence of inflammation, which can cause swelling, pain and—in some instances—loss of mobility. Inflammation happens as a result of the body's immune mechanisms working to prevent harmful stimuli, such as pathogens, harmful cells, or irritants, in the body. It must be noted that inflammation does not mean infection; inflammation is the body's immunovascular response to the stimuli. An infection is the interaction between the microbial event and the reaction of the body's inflammatory defense system. However, the body does not respond with inflammation in all instances of infection, and not all inflammation is a direct result of microbial activity (rheumatoid arthritis is an example of inflammation without infection). There are two types of inflammation: acute and chronic. Acute inflammation is the immediate reaction of the body to bacterial pathogens and injured tissues, which usually resolves after a few days. Chronic inflammation is a more persistent type of inflammation that results from pathogens that do not want to degrade, viral infections, persistent foreign bodies, or autoimmune reactions. Chronic inflammation can persist for many months or even years, causing tissue destruction, fibrosis, and premature cell death. Typically, pharmaceutical drugs, such as NSAIDs, are used to treat the symptoms of both acute and chronic inflammation; however, the side effects of these drugs include but are not limited to kidney problems, stomach problems, and risk of stroke. The aim is to reduce the inflammation that may cause damage without taking in harmful toxins that may lead to additional health issues.

Moringa has been used to treat arthritis and joint pain, and many studies have identified the plant's anti-inflammatory properties. The root, flowers, leaves, and seeds are known to have both analgesic (painkilling) and anti-inflammatory properties. In studies done with Wistar rats and mice, carrageenan-induced edema was injected subcutaneously into the rats' paws after administering a dose of an ethanolic extract of *Moringa*. The studies show that the extract of *Moringa* inhibited the inflammation that would have occurred as a result of the carrageenan. The anti-inflammatory activity of the ethanolic leaf extract of *M. oleifera* may be attributed to its phytochemical ingredients, such as flavonoids, 4-hydroxymellein, beta-sitosterol, and vanillin. Flavonoids play a major role here, as they not only inhibit prostaglandin biosynthesis by inhibiting endoperoxidases but also enzymes like protein kinases and phosphodiesterases that are involved in the inflammation process (Bhattacharya et al. 2014). This study also confirmed the antinociceptive properties of *Moringa*. One of the possible reasons is the inhibition of prostaglandin; prostaglandin usually acts on the peripheral sensory neurons and on central sites within the brain and spinal cord, thus producing pain.

Nutrition: Vitamin and Mineral Deficiency, Energy, Anemia, and *Moringa* as Immunity Booster

Globally, we produce more than we need and discard hundreds of pounds of food daily, yet still there are millions of people around the world who lack the proper nutrition to sustain a healthy lifestyle. Furthermore, the food we produce tends to be packed with sugars and preservatives, which further perpetuates improper nutrition. Finding sources of food that are homegrown and less processed is the crucial ingredient in reducing the epidemic of undernourishment and noncommunicable diseases. Hence the revival of farming in your own backyard; with nutrient rich foods, such as *Moringa*, it is the most logical course of action to increase availability. Aid groups travel around the globe in hopes of educating local populations on how to become self-sustainable as organizations recognize that these at-risk populations should not become dependent on nourishment from international communities. There is always a possibility that the demand may outweigh the supply. Self-sufficiency is now the motto for these communities, which have rich resources in their locally grown plants. In Africa, for example, *Moringa* grows around many homes. However, the true value of the plant's nutritional potential has only come to light in recent years through educational programs in the various villages. *Moringa* long had been used for various ailments, but due to the preparation technique, many of the vital nutrients were being lost.

A barrage of studies has identified the valuable chemical makeup of *Moringa*, which offers very significant quantities of vitamin C, B-complex vitamins, calcium, protein, potassium, magnesium, selenium, zinc, iron, vitamin A, and a good balance of all the essential amino acids. Gram for gram, the nutrients found in *Moringa* exceed those found in spinach, bananas, milk, carrots, and other well-known nutrient-rich foods. The beauty of the *Moringa* leaves specifically is that they can be eaten fresh or cooked or stored as a dry powder for many months without refrigeration.

Price (1985) notes in the Echo Technical Note on *Moringa* that "For a child aged 1–3, a 100g serving of fresh leaves would provide all of the daily requirements of calcium, about 75% of his iron and half his protein needs, as well as important supplies of potassium, B Complex vitamins, coppers and ALL of the essential amino acids. As little as 20g of fresh leaves would provide a child with all the vitamins A and C he needs." He goes on to state that "for pregnant and breastfeeding women … One 100g portion of leaves could provide a woman with over a third of her daily need of calcium and give her important quantities of iron, protein, copper, sulfur and B-vitamins." This vital composition of nutrients and minerals allows the body to (1) process foods in a more efficient manner, thus placing less stress on the body and providing additional energy and (2) build a better defense system against infections, disease, or other unwanted biological invasions. In addition to its immunity- and energy-boosting qualities, the iron content of *Moringa* is significant for protecting against anemia and maintaining normal iron levels. Iron serves multiple roles; it is an essential component of hemoglobin, a type of protein that transfers oxygen from the lungs to the tissues. It is also a component of myoglobin, a protein that provides oxygen to the muscles, and it supports metabolism. It also plays a key role in the synthesis of some hormones and connective tissues. Last but not least, it is vital for overall growth, development, and normal cell functioning. Therefore, incorporating sufficient iron in your diet is crucial.

Male Sexual Enhancer

Sex serves as a key component to any relationship, because it brings two people together in an intimate moment. As they age, men may lose their enthusiasm for sex because of their inability to have or keep an erection during intercourse. Poor health and lifestyle choices, such as a high-stress environment, can also lead to a man's poor performance. Stress level, depression, and obesity all play parts in his ability to perform. This can be a very traumatic experience, and many men seek sexual enhancers to circumvent this problem. Drugs such as Viagra and Cialis were designed to increase the blood flow to the penis. However, the side effects of these types of drugs are endless. On top of that, the drugs are costly. Plant-based aphrodisiacs like *Moringa* are alleged to improve overall sexual performance, particularly for persons who are highly stressed. It is hypothesized that oxidative stress arising from chronic stress leads to the destruction of the interstitial cells of Leydig, which produce testosterone. This decreases the level of serum testosterone and the production of spermatozoa. Also, stress can modulate the neurotransmitters, thus leading to decreased blood flow to the penile area. *Moringa* is a good source of antioxidants and amino acids, which are very beneficial in inhibiting the effects of oxidative stress. Prabsattroo et al. (2015) examined *Moringa* extract's enhancement of sexual performance in stressed rats. In vivo data highlighted that one dose of the medium or low-dose extract increased libido in the rats. Additionally, prolonged treatment with the low dose of the extract enhanced penile erection capacity and increased the number of interstitial cells of Leydig and spermatozoa as well as increasing testosterone levels. The high dose of the extract also increased the number of mountings, interstitial cells of Leydig, and spermatozoa while increasing testosterone levels. Further research needs to be done to determine the underlying factors that allow *Moringa* to work as an aphrodisiac, particularly for persons with stressful lifestyles.

Abortifacient

In India a few years ago, researchers stumbled upon a village where the majority of women used *Moringa* as an effective abortifacient during the early stages of pregnancy. It prompted the researchers to conduct a study utilizing albino rats in the laboratory. The results showed that the experimental group displayed 100 percent abortifacient activity at an average dose of 175 mg of *Moringa* powder (Sethi et al. 1988). That is why it is recommended that women who are pregnant or want to become pregnant should seek advice before consuming any *Moringa* products.

Lactation

In many parts of the Philippines, postpartum mothers use *Moringa* to improve the volume of milk production. It is also believed that the *Moringa* leaves and pods preserve the mother's health and pass on this strength to the nursing child (Price 1985). Estrella et al. (2000) conducted a small double-blind study with a group of women who received either a capsule with 250 mg of *Moringa* or a placebo pill. The women were asked to take the capsule beginning on the third postpartum date, and it was administered every twelve hours. Milk volume was checked on the third, fourth, and fifth days, with a noted increase as the treatment progressed. It must be stated that the long-term effects of the use of *Moringa* as a galactagogue are unknown, and women with hypertension, diabetes, and other illnesses were not used in the study.

Weight Loss

Notions are brewing that *Moringa* aids with weight loss due to its high nutritional content and low fat. The impressive concentration of vitamin B in its leaves aids in digestion and converts our food into energy instead of storing it as fat. However, more research needs to be done to conclusively confirm this claim.

Skin Conditions: Moisturizer, Skin Treatment, and Wound Healing

The Egyptians and Romans used ben oil as a moisturizer for their skin, believing that the oil had healing and protective powers. Fast-forward to the twenty-first century, and we are returning to this plant-based medical wonder. As we stated previously, *Moringa* contains antibacterial and antifungal ingredients that can help to heal skin infections and protect the skin from further invasion. *Moringa* has been used topically to soothe headaches and to treat athlete's foot and other skin infections, warts, and snake bites (Afzal et al. 2012). It also promotes wound healing. After a wound, the body forms blood clots via a complex process called coagulation, which is an important part of hemostasis, which allows for the cessation of blood loss from a damaged blood vessel. The damaged blood vessel is covered by a clot containing platelets and fibrin to stop the bleeding, and then repair on the damaged vessel begins. *Moringa* has anticoagulation capabilities because of the presence of thrombin and plasmin-like enzymes (Satish et al. 2012). Thrombin helps control minor bleeding, while plasmin enzymes present in blood degrade many plasma proteins, including fibrin clots. This confirms *Moringa*'s potential for wound healing.

In addition to its antibacterial and antifungal properties, *Moringa* is a rich source of antioxidants that help to protect against oxidative stress, which can further age the skin, causing wrinkles and lines.

Ultraviolet light from the sun is one of the major contributors to rapid skin damage; it destroys collagen, leading to skin roughness. Aging comes with its own challenges, including loss of elasticity, increased wrinkling, irregular pigmentation, dryness, and skin roughness. One of the avenues that scientists are exploring is the use of natural antioxidants and phenolics found in plant products to counteract this process. *Moringa,* due to its historical use as a skin protector and healer, is being closely examined as an affordable skin cream. To confirm its rejuvenating powers, researchers piloted a study on human subjects using an active cream containing 3 percent concentrated extract of *Moringa* leaves (Atif 2014). Subjects were instructed to apply it on their skin daily for three months, and a video camera with high resolution was used to examine the skin's surface after each month. At the end of the three months, examination showed gradual decreases in roughness, scaliness, and wrinkles for those using the cream with the *Moringa* extract. The improvements noted are believed to be due to the phenolics in *Moringa,* which include kaempferol; quercetin; rutin; and gallic, chlorogenic, ellagic, and ferulic acid. The phenolics, along with the antioxidants vitamins A, C, and B, provide skin protection against enzymes that cause the breakdown of collagen and the elastin in the skin, leading to skin rejuvenation. Vitamin B works as a humectant; that is, it helps to keep the skin moist by attracting water into the *stratum corneum* (the outermost layer of your skin) to soften the skin. Research has shown that beta-carotene has topical photoprotective capabilities by increasing protein and collagen as well as DNA content, thus leading to increased epidermal thickening.

Water Purification

Moringa seeds are one of the best known natural coagulants and serve as a viable replacement for synthetic coagulants. In Malawi, a project was started to run a trial using *Moringa* as the sole coagulant in a large-scale treatment plant (Sutherland et al. 1994). The seeds serve a dual purpose. When the seed is mature, it is approximately

40 percent oil and can be utilized for cooking (Price 1985). The remaining seedcake after oil extraction can serve as soil fertilizer or for the treatment of turbid water. *Moringa* seeds are most effective in waters where there is high turbidity, and they are said to reduce turbidity between 92 and 99 percent. Moreover, the seeds are noted to have water-softening properties as well as being able to correct pH by reducing alkalinity (Anwar et al. 2007). The mechanism for coagulation is that dimeric proteins create a protein powder that is stable and totally soluble in water. The most probable explanation for *Moringa*'s coagulating abilities is adsorption and charge neutralization. In the Sudan, rural women prefer to use the seedcake because of a traditional fear that alum causes gastrointestinal issues and Alzheimer's disease (Anwar et al. 2007).

Waters in some parts of the world contain high concentrations of metals, such as cadmium. Cadmium is a minor metallic element and is one of the natural components in the earth's crust and water; it is present everywhere in our environment. Exposure to certain forms and concentrations of cadmium are known to have a toxic effect on humans. Long-term occupational exposure to cadmium at excessive concentrations can cause adverse health effects on the kidneys and lungs (www.cadmium.org). The seeds of the *Moringa* work as a biosorbent to remove cadmium from water (Sharma et al. 2006). The *Moringa* seed's amino acids constitute a physiologically active group of binding agents that have the ability to interact with metal ions, which leads to an increase of the sorption of metals (Sharma et al. 2006).

Antimicrobial properties have also been reported, and researchers have found that a recombinant protein in the seed has the ability to flocculate Gram-positive and Gram-negative bacterial cells. This causes the microorganisms to clump together and settle in the water, making removal easy. The seeds may also act directly upon the microorganisms, inhibiting their growth either by disrupting cell membranes or by inhibiting essential enzymes. These effects are attributed to the compound 4(α-L-rhamnosyloxy) benzyl isothiocyanate (Eilert, Wolters, and Nadredt 1981). According to ECHO's technical note on the *Moringa* by Price (1985), a unique use for the seed powder is to help with the harvesting of algae, specifically spirulina algae, which is very popular in health foods and cosmetics. The typical process for harvesting algae requires centrifuges, which tends to be a very expensive process. The seed powder, once sprinkled on the water, causes the algae to sink to the bottom, allowing it to be retrieved and dried once the water has been carefully strained.

Ben Oil

Ben oil is the traditional name of the oil produced from the seeds of the *Moringa* tree. Lalas and Tsaknis (2002) noted that the oil contained a high ratio of monounsaturated fatty acids to saturated fatty acids and therefore might be a valuable substitute for highly monounsaturated oils, such as olive oil.

Insect Repellant

One of the uses for *Moringa* is as an all-natural insect repellant. Prabhu et al. (2011) scrutinized *Moringa*'s potential as a larvicidal and repellent against the malarial vector *Anopheles stephensi* Liston. Vector-borne diseases still serve as a major challenge, and we most recently saw the resurgence of dengue and chikungunya in the Caribbean, both borne by mosquitoes. Ongoing use of synthetic insecticides to control the ever-growing mosquito population has led to a disruption in the natural biological control system, the development of mosquito resistance, and undesirable effects on nontarget organisms, including concerns for humans who suffer from respiratory ailments like asthma. The hope is to find a solution in plants, as they are considered a rich source of bioactive chemicals that may replicate their synthetic counterparts. Prabhu's study showed that *Moringa* had a larvicidal effect while being nontoxic to humans. Another benefit, particularly when treating water to prevent larvae growth, is that

Moringa both kills the larvae and purifies the water. Further research will prove *Moringa*'s value as an affordable agent for both repelling and killing mosquito larvae.

Culinary Uses

Fried, baked, boiled, or raw—the *Moringa*'s parts are useful for diverse culinary purposes. The options are seemingly endless when looking to incorporate this nutritional plant into your daily meals, and luckily almost the entire plant is edible. The root is not typically recommended due to the high levels of toxins. It is also best to incorporate the leaves and flowers at the end of the cooking process to prevent the loss of vital nutrients.

Around the world, *Moringa* has been successfully infused in many dishes due to the ease of preparing the various parts. The magnificence of *Moringa* is that the work needed to include its leaves is quite minimal—hence the recent increase in use in many parts of Africa, where malnutrition is exceptionally high. Organizations like the Church World Service have implemented programs to teach locals how to integrate *Moringa* into their diets. Many believe that *Moringa* is "the natural nutrition source for the tropics" (Mahmood et al. 2010). The local people have long eaten the leaves, but they boiled them numerous times and discarded the water after each boiling to remove the bitter taste, ingesting only the remaining nutrient-deficient leaves. They were killing all the essential nutrients that the plant had to offer. The Church World Service and other such organizations developed a campaign to educate local government health workers on the benefits of *Moringa* and the most effective ways to include it in the diets of the local people. Now, instead of boiling out the nutrients, they are using the leaf fresh or as a powder added to sauces.

In some parts of India and Sri Lanka, the drumsticks are boiled until semisoft and united with curries and stews; coconut and mustard often are included in this delicacy. In Cambodia, the leaves are added to soups, such as the traditional *korko* (mixed vegetable soup). In some regions, the flowers are cleansed and combined with gram flour to create *pakoras,* which are deep-fried fritters. In Thai and Indian cooking, the leaves, flowers, and drumsticks are incorporated in curries, stir-fries, salads, soups, omelets. The leaves in particular are said to have a nutty taste when fresh. Some people use the leaves similarly to the way they would use spinach leaves. In the Philippines, *Moringa* leaves are used to create a pesto-like pasta sauce with olive oil and salt. *Lemonsito* is a juice to which *Moringa* can be added to create a cold, icy treat.

The powdered form can also be added to dishes with minimal changes to the taste. When using the powder, only a few tablespoons should be added so as not to change the taste of the dish and to limit overdosing on the nutrients. As the saying goes, too much of a good thing is bad for you.

Additional Uses of the *Moringa* Tree

- With their rapid growth, long taproot, few lateral roots, minimal shade, and production of high-protein biomass, *Moringa* trees are well-suited for use in alley cropping systems.
- *Moringa* leaves provide an excellent material for the production of biogas.
- The wood yields a blue dye that was used in Jamaica and in Senegal.
- *Moringa* trees can serve as a living support for fences around gardens and yards.
- Juice extracted from the leaves can be used to make a foliar nutrient capable of increasing crop yields by up to 30 percent.
- Cultivated intensively and then ploughed back into the soil, green manure can act as a natural fertilizer for other crops.
- The gum produced from a cut *Moringa* tree trunk has been used in calico printing, in making medicines, and as a bland condiment.

- Powdered seeds can be used to clarify honey without boiling. Seed powder can also be used to clarify sugarcane juice.
- *Moringa* flowers are a good source of nectar for honey-producing bees.
- The high bioavailability of *Moringa* leaves and stems makes them an excellent feed for cattle, sheep, goats, pigs, and rabbits.
- In many countries, *Moringa* trees are planted in gardens and along avenues as ornamental trees.
- Incorporating *Moringa* leaves into the soil before planting can prevent damping-off disease *(Pythium debaryanum)* among seedlings.
- The soft, spongy wood makes poor firewood, but the wood pulp is highly suitable for making newsprint and writing paper.
- The bark of the tree can be beaten into a fiber for producing ropes or mats.
- The bark and gum can be used in tanning hides.

Moringa oleifera deserves its claim to fame as the all-natural wonder healer, with numerous health benefits, the majority of which can be vouched for via scientific study. Researchers are working to confirm additional uses for this natural healer, including aiding with the management of herpes simplex virus, boosting immunity for HIV victims, slowing the progression of Alzheimer's disease, treating depression, assisting with eye care, and preventing or treating many other ailments. In a world filled with many man-made toxins that not only pollute our environment but also kill us with various diseases, we need to look for more natural alternatives to slow the rapid degenerative processes that occur not only as we age but as a side effect of our lifestyle choices. Keeping close to the plant has significant benefits to our overall system. Finding the right diet rich in plant products and plant-raised animal products will only serve to compliment *Moringa* supplementation.

As we discussed in this chapter, *Moringa oleifera* is an optimal nutrition source that is cheap and widely available. However, we must remind you of two facts: it is recommended that women who are pregnant should not take *Moringa*, and caution is advised when consuming the root and any of its extracts, as the toxicity levels are notably high. In the next chapter, we will analyze a less known plant called ***Bryophyllum pinnatum*** or, as it is more commonly known, the miracle leaf or life plant.

One must know the normal in order to understand the abnormal.

—Alfred Sparman, MD

8

Bryophyllum Pinnatum

In chapter 7, we touched on a trending medicinal plant, *Moringa oleifera*. We shall continue along this vein with a lesser known diamond in the rough, *Bryophyllum pinnatum* or *Kalanchoe pinnata*. If we conducted another Google search like the one we did for *Moringa*, it would return only about 10 percent as many results. This herb, which is native to Madagascar, has naturalized in many tropical and subtropical regions, such as Asia, Australia, New Zealand, Macaronesia, the Mascarene Islands, the Galapagos Islands, Melanesia, Polynesia, the Philippines, Hawaii, China, Africa, and the West Indies as a weed or an ornamental plant. With the renewed fervor for complementary and alternative medicine, more people are experimenting with medicinal plants to seek natural cures for everyday and chronic ailments.

The advent of the Internet and search engines like Google, Yahoo, and Bing, have sparked an increased awareness of these legendary herbal remedies. A larger percentage of people are purchasing and exploring the applications of medicinal plants than ever before. This renewed interest in complementary and alternative medicine is as a result of people's dissatisfaction with the high costs and potentially hazardous side effects of factory-made pharmaceuticals. The physical evidence of the use of herbal remedies dates back approximately sixty thousand years, and more than a quarter of prescription medicines have been developed from herbs (Halberstein 2005).

B. pinnatum is no stranger to many people throughout the world; however, as is the case with many herbs, many Westerners misunderstand it or are simply unaware of it. Across the globe, from the Caribbean to Africa and Europe, it has been used for ailments either on its own or in combination with other herbs. Now, although the research for *B. pinnatum* is somewhat limited and restricted to nonhuman subjects, the stories of its wondrous work are endless. Fortunately, we have been able to extrapolate positive findings to help you gain a clearer understanding of the true benefits of this plant species.

Plant Description and Composition

B. pinnatum's scientific classification is as follows: the family name is Crassulaceae, and its synonyms include *Bryophyllum calycinum;* the kingdom is Plantae, in the division Magnoliophyta. The class is Magnoliopsida, and the order is Saxifragales. The genus is *Kalanchoe.* The section is *Bryophyllum*, and the species is *K. Pinnata* (Kaur, Bains, and Niazi 2014).

Some of the common names for *B. pinnatum* are maternity plant, *parnabija* (Kamboj and Suluja 2009), air plant, Canterbury bells, cathedral bells, life plant, Mexican love plant, miracle plant, and resurrection plant (Ozolua et al. 2010). The Yoruba in southwest Nigeria call it Odundun or e*we abamoda,* while the Igbo of Nigeria refer to it as *odaa opue,* and for the Chinese it is *da bu si* (Ghasi et al. 2011). Other names for this plant are *zakham-e-hyat, panfutti, ghayamari* (Afzal et al. 2012), Africa never die (Ufelle et al. 2011), and *goodluck* (Tatsimo et al. 2012). In Trinidad and Tobago, its name is wonder of the world (Lans 2006).

B. pinnatum is classed as a common weed that is abundant in crop fields in parts of Africa and is sometimes found in fallow land, especially in forest zones. Although it is classified as a weed, in parts of Australia and the Caribbean, it is grown as an ornamental plant. It is a fleshy herb that branches from the base. The lower and uppermost leaves are simple, while the middle ones are usually pinnately compound with three to five leaflets opposite one another with flat, oval blades with rounded tips. The leaves extend from 5 to 20 cm long and are 2 to 10 cm wide (Ozolua et al. 2010). The leaves are a rich, dark-green color trimmed in red. The plant typically grows to a height of 120 cm. Tiny plantlets occasionally form in the scalloped margins of the leaflets, usually when they become detached from the stems. It is as though a whole new garland of little plants springs from the leaf (Gwehenberger et al. 2004). The plant can also reproduce via seed dispersal. The flowers of *B. pinnatum* grow in branched clusters at the top of its stem and are mainly produced during winter and spring months.

B. pinnatum

Leaf Shoots

The flowers are bell shaped and drooping, measuring up to 7 cm and extending from an individual stalk roughly 10 to 25 cm long. The sepals are prominent and inflated, with a yellowish-green or pale coloring, and are partially fused into a tube that is streaked with pink or reddish blotches. The petals are yellowish-green to dark red and are also partially fused into a tube that divides into four petal lobes near the tip; the petals are approximately 3 to 6 cm in length.

Flowers and Fruit

B. pinnatum's fruit are about 15 mm long with a papery and membranous texture. The fruit are made up of four slender compartments that generally remain enclosed within the old flower parts. The fruit contains numerous minute, slender brownish seeds less than 1 mm long.

Typically, people tend to ingest the leaves, which are said to have an astringent, sour taste, with a sweet postdigestive effect (Afzal et al. 2012). In traditional medicine, the leaves are reported to promote antimicrobial, antiulcer, anti-inflammatory, antifungal, analgesic, and antihypertensive activity (Salahdeen and Yemitan 2006). We did find one study that mentioned the use of the bark for a specific treatment.

Preliminary investigations have found this plant to contain a plethora of phytochemicals, including a new steroidal derivative, stigmast-4, 20 (21), 23-trien-3-one. Numerous studies have identified within the leaf extract the presence of alkaloids, phenols, flavonoids, saponins, tannins, glycosides, sitosterol, anthocyanins, malic acid, quinones, tocopherol, lectins, coumarins, and bufadienolides. In addition, calcium, phosphorous, sodium, potassium malate, magnesium, and trace elements of iron and zinc were noted in the extract. Vitamins were also noted, including ascorbic acid, riboflavin, thiamine and niacin. Researchers also found various enzymes, such as fructose biphosphate aldolase and DNA topoisomerase, which have a role in metabolism. In the extract isolated from the aerial portion, researchers discovered compounds with potent biological elements, such as bersaldegenin-1, 3, 5-orthoacetate, bufadienolide-bryophyllin B and bryophyllin C, bryophyllol, bryophollone, bryophollenone, and bryophynol (Kaur, Bains, and Niazi 2014).

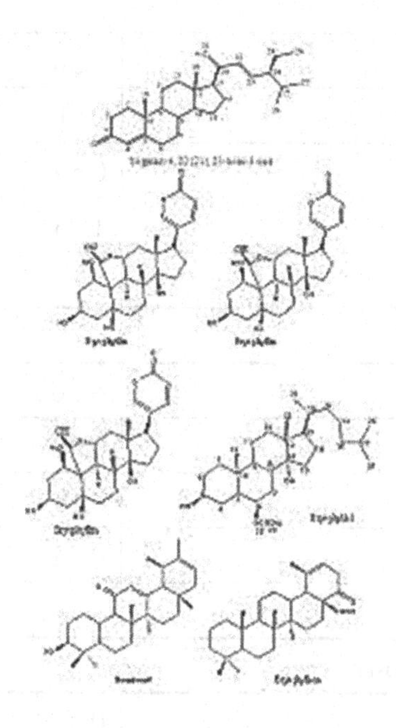

Compounds isolated from *B. pinnatum* (Afzal et al. 2012). The isolation of these phytochemicals and compounds have supported the traditional belief of medical practitioners that the plant extract contains a wealth of curative powers.

Initial research has shown that *Bryophyllum pinnatum*'s nutrient profile is as follows:

Nutrient	**mg/100 g Dry Weight**
Minerals	
Magnesium	0.10
Calcium	0.32
Potassium	0.04
Phosphorous	0.18
Sodium	0.02
Iron	1.85

Zinc	5.38
Manganese	0.51
Copper	0.59
Fiber	1.25
Vitamins	
Ascorbic acid	44.03
Riboflavin	0.42
Thiamine	0.18
Niacin	0.02
Chemical Content	
Carbohydrate	0.429
Protein	1.45
Lipids	2.43
Sugars	
Lactose	0.02
Sucrose	0.49
Glucose	0.26
Fructose	0.23
Maltose	0.56
Other	
Phytochemicals: alkaloids, saponins, tannins, flavonoids, and cardiac glycosides	

Mineral and vitamin content was taken from Okwu and Josiah (2006); chemical content was taken from Alabi, Onibudo, and Amusa (2005).

Bryophyllum pinnatum contains a considerable amount of carbohydrate, crude fiber, protein, fat, ash content, and moisture. Since they contain substantial amounts of nutrients, *Bryophyllum pinnatum* leaves can contribute significantly to the nutrient requirements for healthy growth. The minerals present in the plant also show that *Bryophyllum pinnatum* can serve as a source of nutrient supplements for humans (Nwali et al. 2014).

Historical and Current Uses

In traditional medicine, the leaves of *B. Pinnatum* are reported to have antimicrobial, antiulcer, anti-inflammatory, antifungal, analgesic, and antihypertensive properties (Salahdeen and Yemitan 2006). It is possible to consume the raw leaves or a tea or juice made from the plant, and it also is used in poultices. For example, a poultice for skin ulcers, sprains, and insect stings involves grinding seven fresh leaves and applying them evenly to the affected area twice daily. To create a cough mixture for colds, coughs, and chest congestion, thirteen fresh leaves are warmed over a flame and then rubbed between the palms of the hands until they become juicy. The juice is squeezed into a small pot, which yields approximately six to eight tablespoons of juice. Lime juice and honey are added to the juice and simmered over a low flame for five to seven minutes and then allowed to cool. A tablespoon is administered every three hours for two weeks to help fight infection. These are but two examples extracted from the book *Healing Herbs of Jamaica* by Ivelyn Harris. The Palikur mix the leaf's juice with coconut oil or andiroba oil and then rub it on the forehead for migraines and headaches. Along the Rio Pastaza in Ecuador, the locals use a leaf infusion to help heal broken bones and internal bruises. Indigenous tribes in Peru mix the leaves with *aguardiente* (sugarcane rum) and apply it to their temples to soothe headaches. In Mexico and Nicaragua, it is used to promote menstruation and to assist with childbirth. The plant is also used by the Kerala for treating cancer symptoms (Kamboj and Saluja 2009).

Stories from around the world tell of the healing powers of *B. pinnatum* and its uses in a variety of applications. The leaves of the plant have great value and are used for both internal and external use. Records from as early as 1921 show that it was used as a sedative by psychiatrists in alternative medicine. It was introduced in 1970 at a German complementary and alternative medicine (CAM) center, the Herdecke Community Hospital, as a treatment for premature labor; the clinical outcomes were similar to those of fenoterol (Gwehenberger et al. 2003). In Europe, it was notably used in anthroposophic medicine. In Nigeria, different groups used the plant in a multitude of ways; for example, some southeastern Nigerians used it to facilitate the dropping of the placenta after birth and also applied it on the bodies of young children when they were ill (Afzal et al. 2012). The plant's leaf was mildly exposed to heat and then the juice was extracted and applied to the placenta on a daily basis (Okwu and Josiah 2006). In Benin City, Nigeria, leaves are boiled and filtered through a clean white cloth, and the yield is reconstituted for daily oral use by asthmatic patients (Ozolua et al. 2010). In eastern Nigeria, herbalists use it for the treatment of gonorrhea and genital and mucosal candidiasis (yeast-like fungal infections involving the skin and mucous membrane, including the mouth) (Okwu and Nnamdi 2009). The Igbo in Nigeria believe *B. Pinnatum* has antihypertensive properties; decoctions of the leaf are taken to lower blood pressure. In some instances, the raw leaves of the plant are chewed for this same purpose (Ghasi et al. 2011). In the Mbouda area in Cameroon, researchers surveyed the plants used to treat infectious disease. *B. pinnatum* was one of the most commonly used plants in the area (Tatsimo et al. 2012).

In parts of the Caribbean, a leaf infusion is used to treat dysmenorrhea and high blood pressure and as an antigonorrheal treatment (Halberstein, 2005). In Trinidad, the *Kalanchoe pinnata,* or wonder of the world, is used for cooling, bladder stones, and high blood pressure as well (Lans 2006). In China, the Yao people in Jinping in the Yunnan Province use the plant in the treatment of conditions, including rheumatoid arthritis, stomach bug, injuries from falls, numbness in the limbs, bruises, burns, and ulcers (Ghasi et al. 2011). The plant also is used for hemostasis and wound healing and as a bitter tonic, a carminative, and an analgesic. Ethnopharmacologically, it is used for diarrhea, vomiting, earache, burns, abscesses, gastric ulcers, insect bites, and lithiasis. The juice from the leaves is prescribed for smallpox, coughs, asthma, palpitations, headaches, convulsions, and general debility; the plant is also used in the treatment of edema of the legs (Afzal et al. 2012). The crushed leaves and extracted juice are mixed with shea butter or palm oil and rubbed on abscesses or other swellings and applied to ulcers, burns, and the bodies of young children. The crushed leaves and flowers are used to stop bleeding or for treating wounds and sores, children take an infusion of leaves, and the plant is mixed with clay as a medicine for stomach troubles (Okwu and Josiah 2006). The leaves are used as an astringent, an antiseptic, and a counterirritant against poisonous insect bites. A fresh poultice is applied to sprains, eczema, infections, burns, carbuncle, and erysipelas. In ayurvedic medicine, it is used in vitiated conditions of *vata* and *pitta* for cuts, wounds, hemorrhoids, menorrhagia, boils, external ulcers, scalds, diarrhea, headaches, vomiting, and bronchitis (Raj et al. 2014).

Ethnomedically, it has been used to induce vomiting of blood; to expel *tay tay* worms; and in poultices for head colds, acute and chronic bronchitis, pneumonia, and other forms of respiratory-tract infections (Mudi and Ibrahim 2008). The leaf extract has been routinely used for ailments related to bacterial, fungal, and viral infections; asthma; and kidney stones (Mahata et al. 2012). It also is used in ophthalmic preparations (Ufelle et al. 2011), and the leaves or the whole plant may be used as an analgesic to treat blennorrhea, syphilis, jaundice, dysmenorrhea, convulsions, and dysentery (Tatsimo et al. 2012). Additional uses of note include as a refrigerant, an emollient, a hemostatic remedy, a purifying treatment, a laxative, an anodyne, a carminative, and a disinfectant. It is useful in hematemesis, hemorrhoids, discolorations of the skin, and corns (Afzal et al. 2012). Last but not least, research has shown that *B. pinnatum* is also used to treat heart troubles, epilepsy, arthritis, and whitlow (an infection of the fingertip) (Akinsulire et al. 2007), and the leaf powder is sold as *jakhmehayat* to use as a wound dressing (Afzal et al. 2012).

Research Credence

Much of the research conducted on this plant aims to determine the plausibility of folk practices. Our preliminary research has found support for many of the historical uses of *B. pinnatum.* The future for this herb looks quite promising as demonstrated in the studies below.

Anemia

Anemia is simply a condition in which the blood does not have sufficient red blood cells. It results from a decrease in the production of red blood cells or hemoglobin or from an increase in the loss or destruction of red blood cells. Treatment varies and is dependent on the underlying source of the problem, which can include low iron; medication side effects; or blood loss due to a preexisting condition, such as ulcers. One treatment option for low iron specifically involves iron tablets. Some of the side effects of iron tablets are constipation, diarrhea, nausea, and vomiting. The ideal strategy for supporting persons with anemia would be to introduce a supplement that serves the overall goal of building blood cell count while protecting them from common ailments associated with the condition. Crude methanolic extract of *B. pinnatum* leaves seems to stimulate leukocyte production and hemoglobin synthesis in bone marrow. Ufelle and his team observed that this effect may be a result of the plant's tannin, ascorbic acid, and phenol content. Other phytochemical constituents of *B. pinnatum* that may have affected the hematological parameters in their study include flavonoids, zinc, riboflavin, and niacin. The observed significant increase in hemoglobin concentration and packed cell volume suggests that this crude methanolic leaf extract may have properties that stimulate the bone marrow to produce more hemoglobin when orally administered and may be very useful in the treatment of anemia (Ufelle et al. 2011).

Asthma and Respiratory Infections

Respiratory-tract infections are clinical syndromes that are produced by the inflammation of the trachea, bronchi, and bronchioles caused by bacterial pathogens, such as *Staphylococcus aureus* and *Streptococcus pneumonia*. Despite progress in the development of drugs and antimicrobial agents, drug-resistant microbes and unknown disease-causing microbes pose an enormous public health concern (Mudi and Ibrahim 2008)—hence the need to seek alternative treatments that are safe, efficient, and cost effective. Mudi and Ibrahim partitioned an ethanol extract of *B. pinnatum* into four soluble fractions, subjecting each fraction to antibacterial testing against selected respiratory-tract pathogenic material. One of the fractions in particular, n-hexane, showed activity against the selected microorganisms. This validated the plant's traditional use as a treatment for respiratory infections and provided support for its potential future use as a combatant against respiratory infections.

Furthermore, studies also support *B. pinnatum*'s use in asthma treatment. A study was conducted to investigate the plant's use as a cough remedy and for the prophylaxis of asthma. The results suggested that prophylactic treatment with the extract could reduce the narrowing that follows exposure to spasmogens, lending credibility to the use of the extract for the treatment of asthma in ethnomedicine. However, the exact mechanisms have yet to be pinpointed. Assumptions regarding *B. pinnatum's* of anti-inflammatory properties and flavonoid content play a role in its usage with asthma treatment. Flavonoids have been associated with various smooth muscle relaxant and antioxidant effects. These compounds are known to be very potent antioxidants. They are known to inhibit

Ca^{2+} release and utilization mechanisms in smooth muscles; Ca^{2+} is the forerunner of muscular contractions. The inflammatory process in asthma and airway hyperresponsiveness are often associated with the generation of oxygen radicals (Ozolua et al. 2010).

Histamine and carbachol stimulate H1 and M3 receptors, respectively, on the tracheal smooth muscle cells, culminating in a common pathway for the release of intracellular Ca^{2+}. However, the use of H1-receptor blockers has not been popular in the management of asthma. Afzal et al. note that the methanol extract of the leaves has been reported to act as a histamine H1 receptor antagonist in the ileum, peripheral vasculature, and bronchial muscle (Afzal et al. 2012), so it may aid in the management of asthma. In conclusion, initial research into *B. pinnatum* has found it to possess antispasmodic, antihistaminic, and anti-inflammatory qualities that could be beneficial in developing alternative treatments for asthma patients and to combat respiratory-tract infections.

Allergies

B. pinnatum is believed to exhibit potent antihistamine and antiallergic activity. Reports state that the methanol extract of the leaves acts as a histamine receptor H1 antagonist in the ileum, peripheral vasculature, and bronchial muscle and protects against chemically induced anaphylactic reactions and death by selectively blocking histamine receptors in the lungs (Kamboj and Saluja 2009).

Cancer

Like *M. oleifera, B. pinnatum* has been gaining momentum as a possible protective agent and treatment for cancer patients with minimal side effects. Human papillomavirus (HPV) and cervical cancer go hand in hand. HPV is a very common virus that is transmitted via sexual contact. There are numerous strains of HPV; however, HPV types 16 and 18 have been shown to increase the risk of cervical cancer in women who have persistent infections. The FDA has approved vaccines; however, there has been much controversy surrounding their usage and side effects. Studies have shown that *B. pinnatum* extract has potential as both an anticancer and anti-HPV molecule with possible therapeutic uses against cervical cancer. Mahata et al. conducted a study using *B. pinnatum* leaf extract and human cervical cells to determine the effectiveness of the plant. The leaf extract suppressed the viral transcription of HPV 18 in the cervical cells in conjunction with the inhibition of AP-1 and caused apoptotis. The study further indicated that *B. pinnatum* can act as an anti-HPV molecule and has apoptosis-inducing properties. The study also demonstrated that the antitumor activity of *B. pinnata* is due to its antimutagenic inscription (Mahata et al. 2012).

B. pinnatum is a likely chemoprotective agent. All the bufadienolides found in the plant showed inhibitory activity, and bryophyllin A exhibited the most marked inhibition ability. The flavonoid compounds, which are potent, water-soluble antioxidants and free radical scavengers, prevent oxidative cell damage and therefore promote strong anticancer activity. These results strongly suggest that the bufadienolides of *B. pinnatum* are potential cancer chemoprotective agents (Afzal et al. 2012).

Parasites and Toxins

Depurative agents help to detoxify and purify the body. Usually, the closer to the plant the agent is, the more beneficial. Detoxification may be needed to remove parasitic worms and other parasites within the body. The drugs that can help to expel these nuisances are called anthelmintics; tannins are known to produce anthelmintic activity. Analysis showed that the chloroform methanolic and aqueous extracts of *B. pinnatum* root caused the paralysis and death of worms and showed significant anthelmintic activity (Afzal et al. 2012).

Leishmania

Leishmaniasis is a disease caused by protozoan parasites from more than twenty different *Leishmania* species that are transmitted to humans by the bites of infected female phlebotomine sand flies. In recent years, it has resurfaced due to the high incidence of HIV and is present in thirty-five countries. According to the World Health Organization, one of the biggest threats of the spread of visceral Leishmaniasis is its interaction with HIV-infected persons. This type of leishmaniasis has emerged as an opportunistic infection associated with HIV, and a concomitant HIV infection increases the risk of contracting leishmaniasis more than a hundredfold. The challenge is to find new methods to help protect against this disease. *B. pinnatum*'s flavonoid compounds, which include quercetin and luteolin, have been described as promising antileishmanial drugs with low toxicity (Afzal et al. 2012), which could benefit HIV patients who already have weakened biological systems.

High Blood Pressure

The Yoruba of western Nigeria commonly use the plant in the management of all types and grades of hypertension (Kamboj and Saluja 2009); it is also regularly used in Trinidad and Tobago for this purpose with positive results. In a study, both aqueous and methanolic extracts produced dose-related significant decreases in the arterial blood pressure and heart rates of anesthetized normotensive and hypertensive rats; the hypotensive effects of the extracts were more pronounced in the hypertensive than in the normotensive rats. The extracts also produced significant dose-dependent decreases in the rate and force of contractions of guinea pig isolated atria and inhibited electrical field stimulation as well as potassium and receptor-mediated agonist drug-induced contractions of the rat's isolated thoracic aortic strips in a nonspecific manner. It would appear that cardiac depression and vasodilation contribute significantly to the antihypertensive effect of the herb (Kamboj and Saluja 2009). The calcium present in the plant is a major player in the mechanism that leads to vasodilation. Additional research has shown that the aqueous and methanolic leaf extracts of *B. pinnatum* decrease arterial blood pressures and heart rates of anaesthetized normotensive and hypertensive rats (Afzal et al. 2012). It must also be noted that flavonoids in the intestinal tract help to lower the risk of heart disease (Okwu and Josiah 2006).

Diabetes

In Guatemala and India, the plant is used to treat diabetes (Dewiyanti, Filailla, and Megawati 2012). Hyperglycemia, in particular, is the primary clinical manifestation of diabetes and is thought to contribute to diabetic complications by altering vascular cellular metabolism. Diabetes not only affects sugar levels but also impairs carbohydrate utilization in the diabetic patient, which also leads to accelerated lipolysis which results in elevated plasma triglycerides levels (hyperlipidemia). Consequently, the large amounts of fatty acids available to the liver in diabetic patients lead to excess acetyl coenzyme A (acetyl CoA), which is converted to form ketone bodies, with subsequent damage to the liver. There is an elevation in plasma alanine aminotransferase, which is an indicator of distressed liver function and elevated creatinine levels that result from kidney damage. The increase in the availability of acetyl CoA from the beta-oxidation of fatty acids is also responsible for the subsequent hypercholesterolemia. To reduce the risk of vascular complications in diabetes mellitus, it is necessary to control blood glucose levels, lipid levels, blood pressure, and weight. This is why traditional medical practitioners have managed diabetes for decades with herbal remedies that treat not only the disease but the whole person (Ogbonnia, Odimegwu, and Enwuru 2008).

The plant's aqueous extract caused a significant reduction in the blood glucose level of the fasted normal and fasted streptozotocin induced diabetes mellitus treated rats via a yet-obscure mechanism (Afzal et al. 2012). The presence of zinc in the plants could also play a vital role in the management of diabetes (Kamboj and Saluja 2009). Zinc plays a role in the synthesis, storage, and secretion of insulin and maintains the conformational integrity of insulin in the hexameric form. The quality of the plant's antidiabetic activity is highly dependent on environmental factors. Dewiyanti, Filailla, and Megawati (2012) showed that the antidiabetic activity of *B. pinnatum* ethanolic extract varies; its habitat and sunlight can influence flavonoid compounds in the plant, such as quercetin. Sunlight can increase the quantity of flavonoids in the plant as a result of adaptation induced by the near-wavelength effect of the sunlight. Temperature and water supply at certain seasons can also influence flavonoid constituents in the leaves. According to this study, the drier the climate, the stronger the activity will be (Dewiyanti, Filailla, and Megawati 2012).

Gallstones

The gallbladder is a small sac located on the right-hand side of the body just below the liver. The liver is a hollow organ that concentrates and stores bile. The gallbladder lies in the gallbladder fossa on the inferior aspect of the right lobe of the liver. It has a rounded fundus, a body, and an infundibulum. Bile juice, which is also called gall, is a greenish-brown liquid produced by the liver. The gall goes into the small intestine via the bile ducts to facilitate digestion of fats. *Cholelithiasis* is the medical term for gallstones, which are very common in the UK, the US, and Europe. Gallstones are rare in Africa, China, and Japan. Some believe this may be due to a difference in diet and environmental influences. The motility, concentration, and relaxation of the gallbladder are under the influence of a peptide hormone, cholecystokinin, which is released from the neuroendocrine cells of the duodenum and jejunum. When this bile, which contains high levels of cholesterol, becomes concentrated, hardens, and does not move to the intestines, it becomes gallstones. Gallbladder stones are mainly cholesterol stones, but pigment stones and mixed stones, composed of bile pigments and bile salts, also occur. The gallbladder not only concentrates bile but also acidifies it; failure of acidification may promote calcification of gallstones. There are three major types of stones: (1) pure cholesterol stones, which contain at least 90 percent cholesterol; (2) pigment stones that are either brown or black and contain at least 90 percent bilirubin; and (3) mixed-composition stones composed of varying proportions of cholesterol, bilirubin, and other substances, such as calcium carbonate, calcium phosphate, and calcium palmitate (Raj et al. 2014).

Epidemiological risk factors, such as age, gender, diet, obesity, decreased physical activity, rapid weight loss, and oral contraceptives all play a part in the development of gallstones. Reduced bile salt excretion due to cholesterol-lowering drugs is also a factor. One practitioner stated as early as 1941 that food allergies were a common cause of gallbladder disease (Raj et al. 2014). The treatment for cholelithiasis includes pharmacotherapy and surgery; the latter carries risks for patients. Even though drug therapies exist, they are rarely used due to their adverse side effects and the recurrence of gallstones. *B. pinnatum* shows some promise in this arena, as initial studies have concluded that the plant's extracts have antiurolithic activity and the ability to reduce crystal size (Raj et al. 2014). Further scientific studies will need to be conducted to analyze the feasibility and true potential of *B. pinnatum* as a treatment for gallstones.

Antioxidant Power

As we mentioned earlier in our chapter on antioxidants, the physiological burden of free radicals causes an imbalance between oxidants and antioxidants in the body (Afzal et al. 2012). *B. pinnatum* is a rich source of ascorbic acids, riboflavin, thiamin, and niacin. A lack of ascorbic acid impairs the normal formation of intercellular substances, such as collagen, bone matrix, and tooth dentine, throughout the body (Okwu and Josiah 2006). Extracts from the plant have shown significant antioxidant activity in a dose-dependent manner (Raj et al. 2014). In one study, when scientists used DPPH and nitric oxide radical-scavenging methods to detect oxidative activity in aqueous and alcohol extracts of *B. pinnatum* leaves, they found the leaves to have interesting potential free radical scavenging activity (Afzal et al. 2012).

Wound Healing, Painkilling, and Anti-Inflammatory Properties

Scientific research has validated our wondrous extract's traditional use for wound healing and as a painkiller. The high saponin content justifies use of the extract to stop bleeding and treat wounds. Saponin precipitates and coagulates red blood cells. Some of the characteristics of saponins include the formation of foams in aqueous solutions, hemolytic activity, cholesterol-binding properties, and bitterness. The tannins have astringent properties that hasten the healing of wounds and inflamed mucous membranes. Additionally, calcium is very abundant in *B. pinnatum*; normal extracellular calcium concentrations are necessary for blood coagulation and for the integrity of intercellular cement substances (Okwu and Josiah 2006). Nayak et al. (2010) conducted a study to evaluate the extract's wound-healing activity. They saw increased wound contraction and hydroxyproline content in the extract, which further supports the claims made by traditional healers (Raj et al. 2014). The presence of these compounds explains why traditional healers in southeastern Nigeria often used the herb for the treatment of wounds and burns.

B. pinnatum's additional analgesic and anti-inflammatory properties enhance its ability to heal wounds. According to one investigation, the fluid extract has an anti-inflammatory effect in carrageenan-induced edema rats (Afzal et al. 2012). An additional study further supports the anti-inflammatory hypothesis; the plant extract was used on rats following fresh egg albumin–induced pedal edema. The plant extract significantly inhibited fresh egg albumin–induced acute inflammation. Moreover, another study found that the plant exhibited significant antinociceptive effects (Afzal et al. 2012). The antinociceptive and analgesic properties are a plausible explanation for the painkilling effects experienced by those who use the extract for wound healing. The aqueous extract of *B. pinnatum* demonstrates strong analgesic potency comparable in a time- and dose-dependent manner to a nonsteroidal anti-inflammatory drug. Another study showed that the extract was devoid of severe toxic effects, increased the pain threshold of rats, inhibited or reduced abdominal stretches in mice in a dose-dependent manner, and produced weaker anti-inflammatory activity than aspirin (Igwe et al. via Kamboj and Saluja 2009). Lastly, pure isolated alkaloids and their synthetic derivatives are used as basic medicinal agents not only for their analgesic and antispasmodic properties but also for their bactericidal effects (Okwu and Josiah 2006). Ofokansi reported that *B. pinnatum* was effective in the treatment of typhoid fever and other bacterial infections (Okwu and Josiah 2006). Researchers have also noted that the wound healing exhibited by the extract may be attributable to the presence of steroid glycosides. The analgesic and anti-inflammatory activity of a new steroidal derivative obtained from the leaf extract of the plant may be what gives the herb its ability to heal wounds (Afzal et al. 2012).

Antimicrobial and Antibacterial Abilities

The presence of phenolic compounds indicates that the plant possesses antimicrobial properties. Findings such as this have supported the herb's use in treating the navels of newborn babies, which not only heals rapidly but also prevents the formation of infections (Kamboj and Saluja 2009).

The phenolic compounds undergo oxidation and form phenolate ions or quinine; the phenolate ions are able to scavenge and trap microorganisms (Okwu et al. 2006). Two novel flavonoids, 5 methyl 4, 5, 7 trihydroxyl flavones and 4, 3, 5, 7 tetrahydroxy 5 methyl 5 propenamine anthocyanins, showed potential antimicrobial activities against *Pseudomonas aeruginosa*, *Klebsiella pneumonia*, *Escherichia coli*, *Staphylococcus aureus*, *Candida albicans*, and *Aspergillus niger*. When 60 percent methanolic extract of the *B. pinnatum* leaf was used to inhibit the growth of bacteria at a concentration of 25 mg/mL, it showed antibacterial effects. *Staphylococcus aureus* is a major cause of community- and hospital-associated infections, with an estimated mortality of around 7 to 10 percent. About 77 percent of immune-deficient patients' deaths are attributed to microscopic fungi, such as *Candida* species. Typhoid fever, which is caused by *Salmonella typhi,* also continues to be a serious public health problem in developing countries in general and in Sub-Saharan Africa in particular (Tatsimo et al. 2012). The manipulation of *B. pinnatum* into a pharmaceutical-grade drug shows promise as a cheaper alternative to fight these infections.

Neurosedative, Muscle Relaxant, and Antidepressant Qualities

B. pinnatum has been used since 1921 in traditional medicine as an antipsychotic agent; the plant has neurosedative and antidepressant properties. Treating animals with 50 to 200 mg/kg was found to produce quite a significant decrease in locomotor activity in a dose-dependent manner with no ptosis (drooping of the eyelids). Results indicate a significant loss of coordination, a decrease in muscle tone, significant alterations in behavior, and the potentiation of pentobarbital-induced sleeping time. The aqueous leaf extract acts as a depressant on the central nervous system (CNS) (Kamboj and Saluja 2009). Radford et al. (1986) investigated the possibility that the depressant activity of the aqueous leaf extract could be due to the presence of bufadienolide and other water-soluble constituents of the extract. Another study partially attributed the sedative and CNS-depressant properties of the leaf extract to the herb's ability to increase levels of a specific neurotransmitter in the brain called gamma-aminobutyric acid (Raj et al. 2014). This neurotransmitter induces relaxation, reduces anxiety, and calms nervous activity.

Anticonvulsant Abilities

The extract of *B. pinnatum* was found to reduce seizures induced by pentylenetetrazol, strychnine sulfate, and thiosemicarbazide. Researchers also saw increases in the latency period of seizures and a reduction in the duration of seizures induced by the three convulsive agents (Afzal et al. 2012).

Liver and Kidney Protection

B. pinnatum also has hepatoprotective and nephroprotective qualities; that is, it has the ability to protect both the liver and the kidneys from damage. Various scientists have found *B. pinnatum* to be a very effective hepatoprotective agent, as it significantly lowers the enzymes SGOT, SGPT, SALP, and SBLN; increased levels of these enzymes are sensitive indicators of liver injury. In Bundelkhand, India, the juice from the fresh leaves is used very effectively for the treatment of jaundice. Jaundice sometimes occurs due to a dysfunction of normal metabolism and excretion of bilirubin. The pathology may occur before secretion to the liver, within the liver, or after bilirubin is excreted from the liver. Investigators have reported that the aqueous extract of the leaves of *B. pinnatum* possessed potent nephroprotective activity in gentamicin-induced nephrotoxicity in rats. The plant extract also has been found to act as a diuretic and antiurolithic (Afzal et al. 2012). It is thought that the protective effect on gentamicin-induced nephrotoxicity in rats may involve the herb's antioxidant and oxidative radical-scavenging activities (Kamboj and Saluja 2009).

Antifungal Uses

Okwu and Nnamdi (2011) study showed that isolates from *B. pinnatum* inhibited the pathogenic fungi *Candida albicans* and *Aspergillus niger*. The inhibition of these fungi confirms traditional therapeutic claims for the effective use of this herb in the treatment of skin fungus and inflammation. Inhibition of *Candida albicans* has validated the use of *B. pinnatum* in herbal medicine for the treatment of *Candida* infections (Okwu and Nnamdi 2011). *Candida* infections include oral thrush, vaginal yeast infections, and diaper rash. This yeast normally lives on the skin and in mucous membranes without causing infection but sometimes proliferate due to a variety of factors.

Ulcer and Diarrhea Treatment

Ulcer and diarrhea go hand in hand, as diarrhea is one of the symptoms of stomach ulcers in particular. Diarrhea is a manifestation of many illnesses and can wreak havoc on the biological system if left unattended. Folktales have noted the ability of *B. pinnatum* to heal ulcers and soothe diarrhea. Pal and Chaudhari (1991) revealed that the methanolic fraction of the leaves exhibited significant antiulcer activity; premedication tests in rats revealed that the extract provided significant protective action against gastric lesions induced by aspirin, indomethacin, serotonin, reserpine, stress, and ethanol. There was also a significant enhancement of the healing process of acetic acid–induced chronic gastric lesions in mice. Another study by Adesanwo et al. (2007) showed a radical reduction in basal- and histamine-stimulated gastric lesions in a dose-dependent manner, vindicating the plant's traditional use for ulcer treatment (Kamboj 2009). Investigators further demonstrated that the methanol-soluble fraction of *B. pinnatum* leaf extract inhibited the development of a variety of acute ulcers induced in the stomach and duodenum of rats and guinea pigs (Afzal et al. 2012). Flavonoids provide the anti-inflammatory activity that is useful for the treatment of ulcers (Kamboj and Saluja 2009).

Diarrhea is a menace, as it can culminate in morbidity and even mortality due to the loss of fluids and electrolytes from the body. It ranks second only to lower-respiratory-tract infection as the most common infectious cause of death worldwide. Particularly among children less than five years old, diarrheal disease is a significant cause of death; nearly 2 million children in this age group die of it each year. Causes of diarrhea include infective, immunological, and nutritive factors. The World Health Organization has chosen to place special emphasis on the use of traditional medicines in the control and management of diarrhea. One study examined the effects of the herb's aqueous extract in a castor oil–induced diarrhea model. Castor oil produces diarrheal effects due to its active metabolite ricinoleic acid, activation of adenylate cyclase, and stimulation of prostaglandin E and F formation. Recently, nitric oxide also was found to contribute to the diarrheal effect of castor oil. The aqueous extract of *B. pinnatum* leaves reduced the total number of feces in a dose-dependent manner in castor oil–induced diarrhea. The antidiarrheal effect of the extract may be related to an inhibition of muscle contractility and motility as observed by the decrease in intestinal transit by charcoal meal. Antidiarrheal properties of medicinal plants have been found to be due to the presence of tannins, alkaloids, saponins, flavonoids, sterols, and triterpenoids, all which have been identified in the aqueous extract of these leaves (Sharma, Lahkar, and Lahon 2012).

Pregnancy: Uterine Contractility

B. pinnatum was first used to treat preterm labor in 1970; however, because it has been used nearly exclusively in anthroposophical medicine, there have been only two reports of clinical applications (Plangger et al. 2006). *B. pinnatum* is more effective and has fewer side effects than traditional labor inhibitors in preventing preterm delivery. It is, in fact, no less effective than a beta antagonist but is significantly better tolerated. Plangger et al.'s study confirmed the in vitro relaxant effect of *B. pinnatum* in human myometrium. The researchers drew patients from two Swiss and two German centers. Moreover, the good tolerability exhibited in the study has justified a prospective controlled trial of *B. pinnatum* in the same center. Dose-finding studies and further in vitro investigations are needed in order to clarify the herb's uterine cellular action. Conventional labor inhibitors can have considerable adverse effects, most notably on the cardiovascular systems of both the mother and the child. Thus far, no such effects have been reported for *B. pinnatum*. However, relevant studies have been scientifically inadequate. It must be cautioned that the plant should not be used during pregnancy.

Plant Fungus Treatment

As we move away from pesticides and focus more on organic produce, the use of botanicals in crop protection has gained ground in the world of agriculture as an alternative to the toxic, persistent synthetic compounds usually found in pesticides. In developing countries, the move to botanicals is gaining traction due to several factors; for example, the limited external reserves and poor exchange rates of the currencies of these nations limit the quantities of pesticides they can import. The complete removal of subsidies for synthetic herbicides, insecticides, and fungicides has made them inaccessible to the majority of farmers in many African countries. Lastly, it is no longer a hidden fact that these chemicals contaminate stored food commodities, leaving behind harmful residues, especially when an improper dosage is administered. *B. pinnatum* has shown potential. One study showed that the extracts of both fresh and dry leaves of *B. pinnatum* have a high amount of hydrocyanic acid and oxalic acid, which would make the extracts toxic and poisonous. This suggests that extracts of the plant probably contain chemicals that can prevent plants from attack by pathogens like *Pythium aphanidermatum* (a soil-borne pathogen) and *Sclerotium rolfsii* (a fungus). Following these findings, efforts are underway to test the effects of these extracts on plants in the field to determine their agricultural usefulness (Alabi, Onibudo, and Amusa 2005).

The cardiac glycosides present within *Bryophyllum pinnatum* would explain why it may be beneficial for individuals with weak hearts. Secondly, we should mention the immunosuppressive effects. The fatty acids present in *B. pinnatum* may be responsible for its immunosuppressive effects in vivo. Rossi Bergman et al. showed that the aqueous extract of leaves causes significant inhibition of cell-mediated and humoral immune responses in mice (Kamboj and Saluja 2009). Immunosuppressive agents are not suitable for persons with weak or compromised immune systems.

Bryophyllum pinnatum is the second piece of our life-pill puzzle. Thus far, it fits the mold of the ideal multivitamin. Our preliminary research testifies to the plant's endless possibilities due to the presence of such compounds as flavonoids, saponins, newly discovered steroid derivatives, and calcium. The existence of these compounds in this leaf extract makes it a powerhouse product. We shall now continue to our third life-pill component, the powerful vitamin C.

A great leader takes advice from his council and extracts the good.

—Alfred Sparman, MD

9

Vitamin C

Vitamin C, or ascorbic acid, needs no formal introduction, as its existence has been extensively documented for decades. It plays a vital role within our biological systems. It is well-known that most vitamins cannot be naturally synthesized in the body—hence the need for a proper diet. Vitamin C is a powerful, water-soluble antioxidant that forms the first line of defense under many types of oxidizing conditions (Harrison, Bowman, and Polidori 2014). Vitamin C has been regarded as the purest cure for the common cold. Vitamin C's most historically notable use is to prevent and cure scurvy, the "seaman's sickness." Vitamin C has been linked to prevention and treatment of cancer, wounds, heart disease, diabetes, and a bevy of other chronic diseases. From the 1930s onward, Vitamin C has been affordable and readily available, cementing its position as a household staple.

History and Discovery

Swollen, bleeding gums; bruised nostrils; and small ulcers along the lining of the cheeks are some of the symptoms British physician William Stark recorded in the late 1700s while conducting an experiment on diet and health. He was investigating the commonplace notion that a nutritious regimen needed to be both repetitive and bland. Using himself as the test subject, Stark hoped to prove his theory that a healthy diet could be maintained without becoming boring. According to Price (2015), Stark stated, "I confess it will afford me a singular pleasure if I can prove by experiment that a pleasant and varied diet is equally conducive to health."

Drawing inspiration from his friends and peers, Stark designed his diet after completing some preparatory trials to confirm the time needed fully to digest and pass food. Stark began by reducing his diet to just "bread water and after some time sugar" (Price 2015). This first phase of this experiment lasted for approximately two months. During this time Stark consumed mostly meats and starch; his diet was completely devoid of fresh fruits and vegetables. The journal where he kept a log of his physical state documented his deterioration.

"He was faint and vomiting and spent one memorable night passing-and taking notes on—14 watery stools ... his gums turned black, purple streaks showed up on his shoulder and a 'disagreeable, fetid, yellowish fluid' built up in his mouth, but Stark took only a brief break to let his gums heal before plunging back in" (Price 2015).

He then altered his diet slowly by incrementally adding in additional items like goose meat, boiled beef, milk, and figs. The figs were the only fruit or vegetable he consumed throughout the course of this extremely restricted diet, which lasted for almost a year. At twenty-nine years old, Stark died from the cumulative effects of malnutrition. The culprit was most likely a vitamin C deficiency resulting in scurvy.

According to Carpenter (1988), "If we exclude straight forward famine, scurvy is probably the nutritional deficiency disease that caused the most suffering in recorded history." Although it is most often associated with sailors, historically, scurvy has appeared in many groups separated from fresh produce for long periods of time. The California gold rush, the American Civil War, and Arctic expeditions are just a few examples (Carpenter 1988). The ancient Egyptians and Hippocrates were among the first to provide accounts of the malady. Germans mistakenly believed the disease was infectious. A French explorer was taught a cure for scurvy using pine needles while on his travels in the Americas: "In 1540, a French explorer named Jacques Cartier learned of a remedy for scurvy from the Native Americans of lower Canada, which was prepared by extracting the needles of pine trees with hot water" (Goebel, Wong, and Perry 2013).

But what was scurvy? The relationship between the physical symptoms and a lack of vitamin C was not well understood until after the late 1700s. Human bodies are dependent on a diet that incorporates vitamin C, because we do not synthesize it naturally. In 1617, a physician of the British East India Company, John Woodall, was one of the first to suggest a diet rich in citrus fruits. In his book *The Surgeon's Mate,* he states, "The Lemmons, Limes, Tamarinds, Oranges, and other choice of good helps in the Indies ... do farre exceed any that can be carried tither from England" (Eskind Biomedical Library).

A British Royal Navy physician, James Lind, soon supported Woodland's theory with scientific data. Lind proctored a controlled experiment to test the efficacy of citrus fruits on individuals with scurvy. "In 1747, Lind, an officer and naval surgeon in the British Royal Navy, established the fact that oranges and lemons were effective in curing scurvy. He divided patients into 6 groups of 2 and gave each group a different remedy. Only the group given oranges and lemons recovered" (Eskind Biomedical Library).

It took forty years for Lind to persuade the Royal Navy of the validity of his findings. During that time, he fervently continued his practice of administering fruit juices to thousands of patients with scurvy symptoms and released his life work, *Treatise of the Scurvy.* Lind's book was popular among his peers but still contested. Around this time, British explorer James Cook confided in the experimental scientist and president of the Royal Society, Sir James Pringle, regarding the unregistered antiscorbutic methods he had been using. Cook and Pringle noticed that Cook's voyages, though long, had a surprisingly low number of scurvy cases. This was a result of Cook's acquisition of fresh vegetables and juices (and particularly lemons) throughout his voyages. Cook himself, however, not having the background of a dietician, was not sure of the cause of this success. Paradoxically, because Cook was unable to single out exactly which antiscurvy method had proven effective, he may have delayed the identification of fruit juices as a treatment of scurvy. He believed incorrectly that the lack of scurvy on his ship was due to a fermented malt liquor called wort. "Cook administered sweet wort (an infusion of malt), beer (prepared from an experimental, concentrated malt extract), and spruce beer (prepared mainly from molasses), among many other items, in his attempts to prevent and to cure scurvy. Despite the inconclusive nature of his own experiments, he reported favorably after his second voyage (1772–1775) on the use of wort as an antiscorbutic sea medicine (for which purpose it is now known to be useless)" (Stubbs 2003).

Pringle, on the other hand, assumed that this was due to concentration of the juices, as most of their liquid had evaporated. He requested further testing on purified juices. Pringle and Cook received much praise and recognition for their observations and statements from the navy's Sick and Hurt Board. They gave all the credit to Lind for his instructions to incorporate fresh fruit and juices into one's diet.

"In hindsight the story of how Lind's work was received, entailing a lag of 42 years between his clearly described and experimentally 'proven' treatment and its actual introduction by the relevant authorities seemed to some 'one of the most foolish episodes in the whole history of medical science and practice" (Tröhler 2003).

In 1795, the British navy officially adopted this prescription, and the incidence of scurvy declined dramatically. Lime juice was commonly administered to prevent the disease, and sailors were jokingly nicknamed limeys. In the eighteenth and nineteenth centuries, people used the term *antiscorbutic* to describe foods that offered protection against scurvy. The actual mechanism was not discovered until 1912. "Casimir Funk isolated a concentrate from rice polishings that cured polyneuritis in pigeons. He named the concentrate 'vitamine' because it appeared to be vital to life and because it was probably an amine" (Rosenfield 2007).

Amines are organic compounds or functional groups derived from ammonia that contain nitrogen. When it was discovered that Funk hadn't found amines but something else, the *e* was dropped. In the following years, fat-soluble vitamins A and D were discovered, and in 1930, Hungarian scientist Albert Szent-Györgyi and his team discovered vitamin C. Together they conducted a pivotal experiment using guinea pigs who, like humans, have to obtain vitamin C through exogenous dietary sources. Szent-Györgyi had a fascination with the oxidation process and antioxidants. He knew that whatever substance was capable of reversing the oxidation process was found in the adrenal gland, but it ultimately came from orange and cabbage juice. Szent-Györgyi was determined to extract and identify this mystery substance. After he moved to Cambridge, he was able to obtain one gram of the substance from each variable. In the resulting paper, he initially named the substance *ignose,* which he derived by combining the words *ignosco,* meaning "I don't know," and *glucose.* After much prompting from his editor, he agreed to call it hexuronic acid instead (Hoffer 1989). In 1931, Dr. Szent-Györgyi, after returning to his home in Hungary, showed his partner Dr. Svirbely his crystalline extractions. Svirbely had intimately studied various antiscorbutic factors and was able to verify whether a substance contained vitamin C. Together they began experimenting with it on guinea pigs.

"Svirbely divided the animals into two groups: one that received boiled food (boiling destroys vitamin C) and the other that was fed food enriched with hexuronic acid. The latter group flourished, while the first aggregation of guinea pigs developed scurvy-like symptoms and died. Svirbely and Szent-Györgyi decided hexuronic acid—renamed ascorbic acid to reflect its anti-scurvy properties—was indeed the long sought vitamin C" (Schultz 2015).

In 1937, Szent-Györgyi received a Nobel Prize for his discovery of vitamin C. Up until this point, Szent-Györgyi had obtained his supply of ascorbic acid from animal adrenal glands and orange juice. He had become a follower of the ideology that large doses of vitamin C were "much more important than the tiny vitamin doses supported by nutritionists, dietitians, and biochemists" (Hoffer 1989). The process was tedious and did not grant yield of the substance. One night when his wife prepared a dinner heavily seasoned with paprika, he had an idea. Szent-Györgyi's home town of Szed was popular for its supply of paprika; in fact, it was the world's capital source. Szent-Györgyi hated paprika. While contemplating a way to get out of this unfortunate meal, he thought about how he had never tested paprika for vitamin C. He seized his opportunity and took his paprika to the laboratory. "To his delight, he found it was five to six times as rich in Vitamin C as orange juice. Within one week his institute had isolated three pounds of Vitamin C" (Hoffer 1989).

In 1933, Tadeusz Reichstein synthesized vitamin C. This event initiated the rise of vitamin C's mass production and widespread availability. But before this could happen, there was a problem to solve. To begin the synthesis process, he needed a starting material, the glucose molecule L-sorbose. However, although it was a known substance, it was not available on the market. Reichstein was not deterred; he knew that by using certain bacteria strains, regular sugar-alcohol sorbose could be transformed into L-sorbose (Steffan 2012). After many cultures using mold proved to be failures, Reichstein drew inspiration elsewhere. He placed mixtures of water-diluted sorbitol solution, yeast, and vinegar in six glasses and left outside for a couple of days. When he collected the glasses for analysis, half of them showed no change, while the other half contained a startling surprise! Deposits of white crystals were found to be made up of exactly sugar they had sought out to create, L-sorbose. "Responsible for the transformation was a strain of bacteria that was later to be called Acetobacter suboxydans. In one of the glasses, a dead fruit fly was floating in the liquid. On one of its legs, L-sorbose crystals had grown. Evidently, a colony of precisely this type of bacteria had been on the fly's leg" (Steffan 2012).

The lab immediately cultivated the bacteria, and in no time, L-sorbose was readily available. Now that the hard part was over, aided with the help of his student Oppenauer, Reichstein continued his mission to synthesize vitamin C. "Reichstein was able to continue the process of synthesis, acetylation, and oxidation, until it was possible to produce synthetic vitamin C in a way that had great commercial potential" (Steffan 2012). Reichstein's bulk processing of vitamin C made the product cheap and more attractive socially. Under the company name Redoxon, vitamin C was marketed to the public and is still being sold today.

In the 1970s, there was a cultural explosion of support from both scholars and laymen for vitamin C and its healing properties. Linus Pauling was responsible for this. Considered one of the greatest chemists who ever lived, he released a book entitled *Vitamin C and the Common Cold.* Pauling spoke confidently about vitamin C and was quoted as saying that "the regular ingestion of 1000mg (of Vitamin C) leads to the decreased incidence of colds by about 45% … (and) … to a decrease in total illness by about 60%" (Anderson, Reid, and Beaton 1972). His calculations were based on a double-blind study performed by G. Ritzel involving close to 300 students from Switzerland. He also used support from three other clinical studies, including the work of Wilson and Loh: "Of the 108 subjects, 57 received ascorbic acid (200 mg per day) and 46 received placebo tablets. The investigators state that ascorbic acid significantly reduced the incidence, duration, and severity of symptoms of colds, in comparison with the placebo" (Pauling 1971).

Pauling believed that large doses of vitamin C were necessary to maintain optimal health. The public responded to these claims positively, and a vitamin C craze ensued. What caused this reaction?

First, a strong body of evidence supported Pauling's theory. Second, "the fad for 'natural' or 'organic' food and the vitamin supplements of every variety which has entranced multiplying segments of the populace" swept through that decade and has not since relented (O'Niel 1971).

The vitamin C revolution is documented as one of the most significant events in our health-conscious history. It has led to many scientific breakthroughs and assisted in our cultural understanding of how oxidation works and—most importantly—how it can be stabilized and reversed.

Chemical Structure

The active agent is an enolic form of 3-ketol-L-gulofuranlactone christened ascorbic acid or vitamin C (Cameron, Pauling, and Leibovitz 1979). L-ascorbic acid ($C_6H_8O_6$) is the trivial name of vitamin C, and the chemical name is 2-oxo-L-threo-hexono-1, 4-lactone-2, 3-enediol. L-ascorbic and dehydroascorbic acid are its major dietary forms. Ascorbyl palminate is used in commercial antioxidant preparations; this is the only form of ascorbic acid that is not soluble in water.

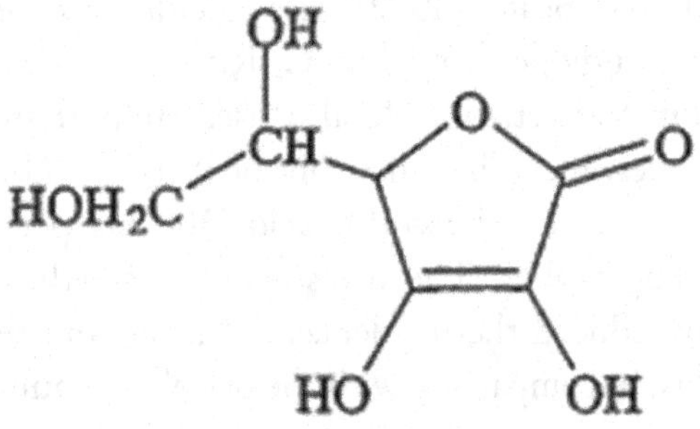

Ascorbic Acid

L-ascorbic acid and its fatty acid esters are used in food additives, antioxidants, browning inhibitors, reducing agents, flavor stabilizers, dough modifiers, and color stabilizers (Naidu 2003). Vitamin C is readily available and easily absorbed via the intestine. At an intake of up 100 mg per day, the body will absorb 80 to 90 percent. However, at higher intake levels, the efficiency of ascorbic acid absorption declines rapidly, and although it is easily absorbed, it is not stored in the body. The average adult has a pool of approximately 1.2 to 2.0 grams of ascorbic acid that can be maintained with an intake of 74 mg per day. Our bodies only need about 140 mg of ascorbic acid per day to saturate the total body pool of vitamin C. The average half-life of this vitamin in an adult human is about ten to twenty days. This is why ascorbic acid has to be regularly replenished through diet or tablets (Naidu 2003).

How Is It Broken Down in the Body?

The major metabolites of ascorbic acid in humans are dehydroascorbic acid; 2, 3-diketogulonic acid; and oxalic acid. The main route of elimination of ascorbic acid and its metabolites is in urine. It is excreted unchanged when high doses of ascorbic acid are consumed (Naidu 2003).

Sources

As we are well aware, sources of vitamin C are typically fruits and vegetables; however, some fruits and vegetables contain significantly more of this vitamin than others. The table below presents some of the items with the milligrams present based on its serving size.

Food	Serving Size	Vitamin C (mg)
Vegetables		
Peppers (red, yellow), raw	125 mL (½ cup)	101–144

Peppers (red, green), cooked	125 mL (½ cup)	121–132
Peppers (green), raw	125 mL (½ cup)	63
Broccoli, cooked	125 mL (½ cup)	54
Cabbage (red), raw	250 mL (1 cup)	54
Brussels sprouts, cooked	125 mL (4 sprouts)	38–52
Kohlrabi, cooked	125 mL (½ cup)	47
Broccoli, raw	125 mL (½ cup)	42
Snow peas, cooked	125 mL (½ cup)	41
Cabbage, cooked	125 mL (½ cup)	30
Cauliflower, raw or cooked	125 mL (½ cup)	26–29
Kale, cooked	125 mL (½ cup)	28
Rapini, cooked	125 mL (½ cup)	24
Potato, with skin, cooked	1 medium	17–24
Bok choy, cooked	125 mL (1/2 cup)	23
Sweet potato, with skin, cooked	1 medium	22
Asparagus, frozen, cooked	6 spears	22
Balsam pear/bitter melon	125 mL (½ cup)	22
Turnip greens, cooked	125 mL (½ cup)	21
Snow peas, raw	125 mL (½ cup)	20
Collards, cooked	125 mL (½ cup)	18
Tomato, raw	1 medium	16
Tomato sauce, canned	125 mL (½ cup)	15
Fruit		
Guava	1 fruit	206
Papaya	½ fruit	94

Kiwi	1 large	84
Orange	1 medium	59–83
Lychee	10 fruits	69
Strawberries	125 mL (½ cup)	52
Pineapple	125 mL (½ cup)	39–49
Grapefruit (pink or red)	½ fruit	38–47
Clementine	1 fruit	36
Cantaloupe	125 mL (½ cup)	31
Mango	½ fruit	29
Avocado, Florida	½ fruit	26
Soursop	125 mL (½ cup)	25
Tangerine or mandarin	1 medium	22
Persimmon	125 mL (½ cup)	17
Berries (raspberries, blueberries, or blackberries)	125 mL (½ cup)	14–17
Juice		
Juice (orange, grapefruit, apple, pineapple, or grape), vitamin C added	125 mL (½ cup)	23–66
Fruit and vegetable cocktail	125 mL (½ cup)	35–40
Guava nectar	125 mL (½ cup)	26
Grain Products	This food group contains very little of this nutrient.	
Milk and Alternatives	This food group contains very little of this nutrient.	
Meats and Alternatives	This food group contains very little of this nutrient.	

Vitamin C content taken from the Dietitians of Canada (http://www.dietitians.ca/Your-Health/Nutrition-A-Z/Vitamins/Food-Sources-of-Vitamin-C.aspx).

Biological Significance and Role in Human Function

Most plant and animal species have the ability to synthesize vitamin C from glucose and galactose through the uronic acid pathway, but primates cannot do so because of a deficiency of the enzyme gulonolactone oxidase (Chambial et al. 2013). The bioavailability of vitamin C essentially depends on its absorption from the intestine and renal excretion. It is consumed either in food or dietary supplements. It is then either absorbed by the epithelial cells of the small intestine by SVCT1 (sodium-dependent vitamin C cotransporter), a key protein for vitamin C uptake, or diffuses into the surrounding capillaries and then the circulatory system. Together, intestinal absorption and renal excretion control the serum level of vitamin C and thus its bioavailability. At low concentrations, most vitamin C is absorbed in the small intestine and reabsorbed from the renal tubule. However, at high concentrations, SVCT1 becomes down regulated, which limits the amount of ascorbic acid absorbed from the intestine and kidneys. This creates a physiological restriction on the maximal effective serum vitamin C concentration attainable by oral consumption. In the event of high consumption levels, ascorbic acid and its metabolites, such as dehydroascorbic acid, 2, 3-diketogulonic acid and oxalic acid, are removed via the kidneys in humans (Chambial et al. 2013).

The physiological functions of ascorbic acid are largely dependent on the oxidation-reduction properties of this vitamin, as it is a well-known electron donor (Naidu 2003). The body requires vitamin C for normal physiological functions. It helps in the synthesis and metabolism of tyrosine, folic acid, and tryptophan and the hydroxylation of glycine, proline, lysine carnitine, and catecholamine. It facilitates the conversion of cholesterol into bile acids and hence lowers blood cholesterol. Also, it increases the absorption of iron in the gut by reducing ferric iron to ferrous iron. As an antioxidant, it protects the body from various deleterious effects of free radicals, pollutants, and toxins. Reduced bioavailability of vitamin C is often associated with stress; smoking; fever; viral illnesses; use of antibiotics, painkillers, and alcohol; and exposure to petroleum products, carbon monoxide, heavy metals, and other environmental hazards. However, the precise mechanisms that cause low vitamin C levels in the body are not well defined (Chambial et al. 2013).

Vitamin C as an Enzyme

Vitamin C as an enzyme has five major tasks in our bodies. These biochemical functions largely depend on the oxidation-reduction properties of L-ascorbic acid, which is a cofactor for hydroxylation and the activity of monooxygenase enzymes in the synthesis of collagen, carnitine, and neurotransmitters. Ascorbic acid accelerates hydroxylation reactions by maintaining the active center of metal ions in a reduced state for optimal activity of the enzymes hydroxylase and oxygenase; thus, it is crucial in the maintenance of collagen. Collagen represents about one-third of total body protein and constitutes the major protein of our skin, bones, teeth, cartilage, tendons, blood vessels, heart valves, intervertebral discs, cornea, and eye lens.

Chambial et al. (2013) highlighted the major biochemical functions of ascorbic acid. It is essential for the maintenance of the enzymes prolyl and lysyl hydroxylase in an active form; when inadequate ascorbic acid is present, a reduction in hydroxylation of proline and lysine occurs, affecting collagen synthesis. Hydroxylation, simply put, is the chemical process that introduces a hydroxyl group or an organic compound. Ascorbic acid also is a key cofactor for hydroxylations involved in the synthesis of muscle carnitine and beta-hydroxybutyric acid. Carnitine is needed for the transportation of long-chain fatty acids into the mitochondria for the production of energy. Ascorbic acid is also required as a cofactor for the enzyme dopamine-beta-hydroxylase, which catalyzes the conversion of the neurotransmitter, dopamine, to norepinephrine. This conversion is essential for the synthesis of catecholamines. Catecholamines are key neurotransmitters in the central nervous system and act as hormones in the blood circulatory system. Ascorbic acid catalyzes other enzymatic reactions involving amidation necessary for maximal activity of hormones like oxytocin and vasopressin.

Ascorbic acid is also involved in the transformation of cholesterol to bile acid, as it modulates the microsomal 7alpha-hydroxylation, the rate-limiting reaction of cholesterol catabolism in the liver. A deficiency of ascorbic acid will affect this conversion, resulting in the accumulation of cholesterol in the liver, which leads to such complications as hypercholesterolemia and the formation of gallstones (Chambial et al. 2013).

Vitamin C and the Immune System

Vitamin C affects several components of the immune system. It appears to play a role in a number of neutrophil functions, including increasing the movement of chemically responsive organelles and particulate ingestion. It also protects against the effects of free radicals (particularly superoxide anions), suppresses highly reactive neutrophil enzymes, heightens lysozyme-regulated nonoxidative killing, and stimulates metabolic pathways. It seems that vitamin C tends to be more involved in cell-mediated immune responses than in humoral immunity, which consists of an antibody-regulated defense conducted by extracellular fluids. This can be seen through decline in the accelerated response of T cells (white blood cells used during cell-mediated response) of patients with Crohn's disease given vitamin C supplements. Studies also show that vitamin C is a cofactor of other macronutrients, forming a synergistic relationship that supports the skin's barrier function in addition to other immune cell activities. The following are some observations regarding the immune function of vitamin C:

- Vitamin C, along with other micronutrients, helps to manage potential damage caused by free radicals at the cellular level and modulates immune-cell functions through the regulation of redox-sensitive transcription factors. It also affects production of cytokines and prostaglandins. Adequate intake of vitamin C and other vitamins and micronutrients like vitamin B6, folate, vitamin B12, vitamin E, selenium, zinc, copper, and iron supports a Th1 cytokine-mediated immune response with sufficient production of proinflammatory cytokines.
- Vitamin C inhibits the excessive activation of the immune system to prevent tissue damage. It also supports antibacterial activity and stimulates natural killer cells.
- Vitamin C modulates synthesis of proinflammatory cytokines or expression of adhesive molecules.
- Vitamin C contributes in maintenance of the redox integrity of cells and thereby protects them against ROS generated during the respiratory burst and inflammatory response.
- Vitamin C has diverse roles as an antioxidant, protecting the immune cells against intracellular ROS production during inflammatory response, acting as an enzymatic cofactor, and maintaining tissue integrity. It plays a crucial role in the formation of skin, epithelial, and endothelial barriers. Recently, vitamin C supplementation has been found to be beneficial in various inflammatory conditions. (Chambial et al. 2013)

Vitamin C and Tissue Healing

Wound healing requires the presence of collagen to provide a fibrous cross-linking to create new toughness in damaged tissue. It has been reported that vitamin C supplementation expedites this process. The guinea pig was used in a study to confirm that after supplementation with the vitamin, scar tissue achieved maximum tensile strength. Numerous studies validate this claim. Many have concluded that due to ascorbic acid's ability to stimulate collagen synthesis, it significantly improves wound repair time. However, adequate supplies of ascorbic acid need to be available in the body pool to facilitate this, especially for those who are undergoing procedures, as large amounts of vitamin C are needed during postoperative healing to allow for collagen synthesis. Researchers recently proved that irradiated wounds healed more effectively when a pretreatment with ascorbic acid was administered. It therefore has been suggested that vitamin C be included in a patient's care plan to accelerate wound repair (Chambial et al. 2013).

Deficiencies

A survey showed that 22 percent of US adults have below adequate vitamin C levels, and about 6 percent of the adult population is classified as vitamin C deficient. Nonsupplementing men between the ages of twenty and forty-nine are particularly at risk for poor vitamin C status. Although overt scurvy is reported occasionally in medical literature, poor vitamin C status usually is undiagnosed, as early symptoms are nonspecific and unremarkable. Some of the symptoms typically include fatigue, malaise, depression, and irritability. The fatigue and weakness of early scurvy can be attributed to defective carnitine production and the resultant reduction in fat oxidation. Carnitine is responsible for energy production. Furthermore, vitamin C is a cosubstrate for the dopamine enzyme, which converts the neurotransmitter dopamine to norepinephrine. The results of inefficient neurotransmitter conversions may be responsible for the depression and mood swings that are typical of early scurvy (Johnston, Barkyoumb, and Schumacher 2014).

Scurvy is the disease associated with ascorbic acid deficiency. It is characterized by spongy, swollen, bleeding gums; dry skin; open sores; fatigue; impaired wound healing; and depression. Scurvy is relatively rare nowadays due to adequate intake of ascorbic acid through fresh vegetables, fruits, and supplements.

Cancer

A cancer patient's ability to fight the irregular cells that are growing in his or her body depends partially on the quality of his or her immune system. Its ability to ward off further infections is very important to the patient's cancer management and overall prognosis. The idea behind including vitamin C as part of the cancer treatment process is to enhance the body's resistance. This train of thought has been quite controversial historically; however, some research is encouraging. Vitamin C was first proposed as a possible form of treatment in 1949. Cameron, Pauling, and Leibovitz noted that high doses of vitamin C improved the survival of terminally ill cancer patients. Pauling and Cameron followed up this research with a study in 1970 involving one hundred cancer patients. Each patient from the group received 10 g of vitamin C and the results compared to a group of a thousand cancer patients who were receiving traditional forms of treatment. It was found that 10.3 percent of the patients who had received vitamin C survived, while 100 percent from the group that received conventional therapy died. This study sparked other researchers to search for the mechanisms that improved the survival odds for the vitamin C recipients.

One of the possible mechanisms relates to free radicals. Free radicals are said to be to blame for the DNA damage that initiates tumor growth. It is believed that vitamin C is able to quell free radicals before they can cause damage or work as prooxidants, feeding the body's own free radicals to destroy tumors in their early stages (Naidu 2003). As researchers continued to gain a better understanding of vitamin C's role in cancer care, they discovered that vitamin C killed only some cancer cells. One study sought to understand whether pharmacological-grade ascorbic acid selectively affected the survival of cancer cells (Chen et al. 2005). The researchers studied nine cancer lines along with four normal cell types. The data showed that the ascorbic acid only killed some of the cancer cells but did not harm the normal cells. It seems the ascorbic acid destroyed the cancer cells due to the protein-dependent extracellular generation of H_2O_2. It is unclear why the H_2O_2 killed some of the cancer cells but not any of the normal ones. Another supporting benefit of intravenous vitamin C specifically is that it can enhance the immune system and build supportive collagen, aiding in the treatment and prevention of cancer.

A high intake of ascorbic acid or foods rich in vitamin C has been found to create a significant reduction in the risk of stomach cancer. Biological and physiological evidence suggests that the free radical–scavenging abilities of ascorbic acid inhibit the formation of potentially carcinogenic N-nitroso compounds, protecting the stomach from cancer. Ovarian cancer was found to be inhibited by ascorbyl strearate, an offshoot of ascorbic acid, which halted carcinogenic cell proliferation and caused cell death to ovarian cancer cells. Although the etiology of Vitamin C and cancer is yet to be understood, overwhelming evidence is accumulating that demonstrates a connection between high intake of vitamin C and a reduced risk of esophageal, oral cavity, stomach, pancreatic, cervical, rectal, breast, and nonhormonal cancers. Independent evidence from various committees and researchers supports the diet's role in improving risk factors. Research suggests the importance of a diet rich in fruits and vegetables in reducing the risk of different types of cancers and decreasing the cancer mortality rate. Although the mechanism of vitamin C still eludes researchers, many complementary and alternative practitioners worldwide currently use high doses of intravenous ascorbate to treat their patients. Incorporation of ascorbic acid in cancer treatment will do more good than harm, as one of the most easily controlled elements of cancer risk is diet.

Cardiovascular Disease

Increasing intake of antioxidants is associated with a reduced risk of coronary artery disease. However the exact mechanisms remain unclear. Multiple mechanisms have been suggested, such as preservation of normal endothelial function by decreasing oxidative stress; increased oxidative stress may impair the endothelium-derived relaxing factor (EDRF), preventing smooth muscle relaxation (Levine et al. 1996). Ascorbic acid plays an important role in the regulation of intracellular redox reaction through its interaction with glutathione. Glutathione is a vital antioxidant that helps prevent damage from ROS to important cellular components. Under conditions of increased oxidative stress, glutathione is oxidized to glutathione disulfide, which can have a negative impact on the body's immune responses. The presence of ascorbic acid protects glutathione from oxidization and improves EDRF action via the synthesis of nitrogen oxide, which aids smooth muscle contractions. It is important to note that the loss of flow-mediated dilation within the endothelium has been implicated in the pathophysiology of abnormal coronary reactivity, which is a stimulant for mental stress and exercise. A placebo-controlled study demonstrated that oral intake of 2 g of ascorbic acid was able to reverse endothelial vasomotor dysfunction in patients with coronary artery disease (Levine et al. 1996). This confirms ascorbic acid's effectiveness at restoring endothelial release and EDRF action in the brachial artery among patients with coronary artery disease.

Atherosclerosis

The buildup of fats, cholesterol, and other substances within our arterial walls is a typical occurrence as we age. This occurrence is also known as atherosclerosis. A 1947 study discovered the significance of vitamin C deficiency for the etiology of both atherosclerosis and dyslipidemia. The study highlighted the capabilities of vitamin C to lower cholesterol levels in hypercholesterolemic patients. A deficiency in vitamin C led to the accumulation of cholesterol in the thoracic aorta as well as changes to the blood vessels. Links also have also been noted with LDL and its impact on atherogenesis. It has been reported that lipid peroxidation and oxidative modification of LDL may lead to atherosclerosis. As an antioxidant scavenger of free radicals, ascorbic acid has the ability to prevent the oxidation of LDL molecules, thus reducing oxidative stress and protecting against

LDL oxidative changes. Marc, Kothari, and Sharma observed that vitamin C intake caused significant reductions in LDL presence and increased HDL levels, decreasing the risk of coronary artery disease (Chambial et al. 2013).

Vitamin C's ability to lower cholesterol can be inhibited depending on initial cholesterol levels, age, and sex as well as the method of administration and the quantity of the dose. Age is a principal factor, as older persons tend to have lower body pools of vitamin C. Therefore, they may be more responsive to the intake of vitamin C than younger individuals. Vitamin C shows great potential as an affordable treatment that can be combined with more conventional medical interventions. Based on the research compiled thus far, vitamin C is able to strengthen artery walls via collagen synthesis and inhibiting the unwanted adhesion of white blood cells to damaged arteries. It prevents endothelial dysfunction, thus providing adequate protection against many vascular inflammatory conditions and atherosclerosis (Chambial et al. 2013).

Chronic Smokers and Improved Endothelial Function

Chronic smoking is the leading cause of many noncommunicable diseases, including coronary and peripheral vascular disease. The large number of oxidants present in cigarette smoke lead to oxidative damage to critical biological substances. The smoke produced is known to contain large amounts of free radicals and prooxidants, such as transition metals and nitrogen dioxide. The particulate phase produces more negative substances, including high concentrations of lipophilic quinones, which produce oxidants like O_2^- and H_2O_2 during oxidation redox cycling.

Reports indicate that chronic smokers tend to present with abnormal endothelial function that is due either to damage to the inner lining of blood vessels or to an irregularity in their ability to pump blood efficiently through the body. It is therefore hypothesized that endothelial dysfunction is a key factor in the initiation of vascular disease. Smokers also have unusually low levels of vitamin C in their plasma, which may exacerbate their endothelial deterioration. Additionally, in smokers, there seems to be an imbalance in the vasculature between antioxidants and prooxidants, which leads to the negative formation of oxidized LDL, compounding the smoker's dilemma. Smokers' low vitamin C levels are partly responsible for the endothelial dysfunction exhibited. Heitzer, Just, and Münzel conducted a study that demonstrated that impaired endothelium-dependent vasodilation in chronic smokers dramatically improved with the combined administration of vitamin C and other traditional treatment methods. Their findings suggest that increased vitamin C will decrease oxidative stress. The thought is that the vitamin's antioxidant properties help to quench the free radicals produced within the blood vessels. According to these findings, acute treatment with vitamin C could almost reverse endothelial dysfunction in chronic smokers (Heitzer, Just, and Münzel 1996).

Diabetes

As we mentioned in chapter 7, diabetes mellitus patients are at risk for a multitude of dysfunctions. Diabetic retinopathy—that is, damage to the eyes—is one that affects many patients. Additionally, the kidneys, nerves, heart, and blood vessels can all be affected by poorly managed hyperglycemia. This poor management causes a chain reaction by activating oxidative stress and particularly the proliferation of ROS. A longitudinal study called the Norfolk Prospective Study demonstrated the link between fruit and vegetable intake and vitamin C plasma levels in relation to type 2 diabetes. Following 21,000 participants over twelve years later, the study identified 735 incident cases of diabetes. An inverse relationship was discovered between risk of diabetes and plasma levels of vitamin C. Studies such as this have provided the evidence needed to cement the relationship between vitamin C intake and decreased risk of developing diabetes mellitus. Chambial et al. (2013) conducted a study to further observe vitamin C levels in diabetics; they noticed lower levels in these subjects. They also found that vitamin C was associated with components of metabolic syndrome; as a metabolic component became active, there was a significant reduction in vitamin C levels in the body.

Clinical evidence has shown that diabetes is closely associated with oxidative stress. The increased generation of ROS leads to increases in the presence of both type 1 and type 2 diabetes. Hyperglycemia and the presence of ROS present further complications for the diabetic patient's vascular system. One of the most important microvascular complications is diabetic nephropathy—that is, kidney damage. Hyperglycemia is responsible for the generation of ROS within the microvascular structure, leading to endothelial dysfunction. Vitamin C is able to impair the hyperglycemic reaction and protect the endothelial lining of diabetics. In recent studies, supplementation with vitamins C and E has been found to relieve oxidative stress in the blood and tissues of diabetic aged rats by tempering the antioxidant and lipid profiles. Supplementation with these two vitamins has also been shown to prevent oxidative stress–induced retinopathy by reducing neovascularization and inhibiting other negative activities that lead to damage of the blood vessels in the eye.

Acute Care

Pauling (1971) states in his book review "Vitamin C and the Common Cold" that "the regular ingestion of about 200 mg of ascorbic acid per day leads to a decrease in incidence of colds by about 15% and regular ingestion of 1000 mg a day leads to the decreased incidence of colds by about 45%. Moreover, these quantities of ascorbic acid ingested are indicated by the reported results that lead to a decrease in total illness (integrated morbidity) by about 30% and 60% respectively." He was one of the first to acknowledge the potential therapeutic role of vitamin C in flu prevention and to introduce the concept of high-dose vitamin C treatment for acute care. Large doses have now become common practice as a treatment tool for disorders, including diabetes, atherosclerosis, the common cold, cataracts, glaucoma, macular degeneration, stroke, heart disease, and cancer.

During an infection, T cells are activated to destroy infected targets using a process known as lysing, which involves the production of large quantities of cytokines. T cells also assist B cells to produce immunoglobulins in order to hinder the inflammatory process. Ascorbic acid supports this defense via the stimulation of the T cells, which it protects from apoptosis, thus allowing for the proliferation of T cells to fight the infection. Researchers have proposed that this scenario is responsible for the enhanced response seen after administering vitamin C to persons suffering from a runny or congested nose. Clinical trials have clearly demonstrated a reduction in the severity and duration of colds during the infection stage after vitamin C has been administered (Chambial et al. 2013).

Men's Fertility

Ascorbic acid is the main antioxidant in the seminal plasma of fertile men. Its concentration in seminal plasma is almost ten times higher than the blood plasma concentration (Colagar et al. 2009). Various studies have found significant differences in ascorbic acid content in the seminal plasma of fertile and infertile men. Furthermore, the percentage of sperm with normal morphology correlated with high seminal ascorbic acid in both groups. Ascorbic acid deficiency may lead to an increased presence of ROS, leading to oxidative damage; the increase in ROS was observed in the semen of 25 to 45 percent of men who were infertile. Others have also observed deleterious effects oxidative stress on male fertility. Further studies report that supplementation of ascorbic acid leads to reduction in ROS concentration, reduction in sperm DNA oxidation, and increased sperm quality. Vitamin C supplementation in men may help to improve sperm quality (Chamberlain et al. 2013).

Pregnancy

In some developing countries, pregnant women are frequently hospitalized for preventable ailments, such as anemia, particularly iron-deficient anemia, and respiratory-tract infections. It has been estimated that 33 percent to 75 percent of women in developing countries are affected by anemia in pregnancy. In a randomized study by Hans and Edwards, daily vitamin C supplementation resulted in positive clinical signs for the group of pregnant women treated, with no reported hospitalizations during their pregnancies compared to the control group, who received only the routine supplements for prevention of anemia. These findings can be interpreted in two ways. First, there may have been an increase in gastric absorption of iron in the presence of ascorbic acid. Second, the presence of high amounts of ascorbic acid may have protected red blood cells from oxidation (Hans and Edwards 2010).

Anemia

Ascorbic acid enhances the availability and dietary absorption of iron from nonheme sources (that is, iron from plants). The reduction of iron by ascorbic acid has been suggested to increase the absorption of nonheme iron in the body. Fruits rich in vitamin C are said to increase the bioavailability of iron from staple cereals and pulses. Recent observations suggest that vitamin C, as a result of improving the bioavailability of iron, aids in protecting against anemia.

Neurodegenerative Support

Schizophrenia is a debilitating, relatively common major neurological disorder that creates a significant economic burden. As it is a multifactorial disease, the outcomes are typically poor even with access to the best available treatments. Vitamin C first captured the attention of psychiatrists as a part of treatment for schizophrenia almost seven decades ago. A study conducted on twelve schizophrenics showed that intravenous injection of a large dose of vitamin C produced improvement in mental condition in 75 percent of the patients. This improvement may relate to the increased generation of free radicals in the pathogenesis of schizophrenia. Also, alterations in the optimal activities of antioxidant enzymes and lipid peroxidation in the blood have been detected in schizophrenics. Another study suggests that antioxidant supplement therapy, including high doses of vitamin C, may be useful as a complementary therapy to improve outcomes for patients with stress-induced psychiatric disorders. Vitamin C may also be beneficial for those suffering neurodegenerative disorders like Alzheimer's disease. Evidence supports the maintenance of a healthy vitamin C level as a protection against age-related cognitive decline (Chambial et al. 2013).

Kidney Disease

Vitamin C has been shown to reduce renal damage caused by a variety of insults, such as postischemic stress, some cancer drugs, antibacterial antibiotics, and potassium bromate, in animals. Spargias et al. conducted a study to determine vitamin C's ability to protect nephrology patients who had to undergo cardiac procedures using contrast medium, which tends to cause further damage to their already compromised systems. The study demonstrated that ascorbic acid is a safe, well-tolerated, inexpensive, and readily available oral antioxidant that appears to prevent complications related to contrast-mediated nephropathy after invasive coronary-imaging procedures in these patients. These findings are consistent with the hypothesis that contrast-mediated nephropathy is caused in whole or in part by oxidative stress. Vitamin C also can potentially protect patients with renal dysfunction undergoing cardiac contrast-dye procedures (Spargias et al. 2004).

Critically Ill Patients

Critically ill patients whose systems are severely compromised require additional supplementation to foster their recuperation. "Pre-clinical studies show that high-dose vitamin C can prevent or restore ROS-induced microcirculatory flow impairment, prevent or restore vascular responsiveness to vasoconstrictors, preserve endothelial barrier and augment antibacterial defense. These protective effects against oxidative stress seem to mitigate organ injury and dysfunction, and promote recovery in most clinical studies after cardiac revascularization and in critically ill patients" (Straaten, Spoelstra-de Man, and de Waard 2014).

Cardiac Surgical Patients

Vitamin C is a valuable tool when administered before surgery in order to help reduce the incidence of new postoperative atrial fibrillation in patients undergoing coronary artery bypass. A recent study was able to replicate the results of an older Chinese study that found using high doses of intravenous vitamin C caused a reduction in cardiac injury. It also improved cardiac performance and led to a shorter overall patient stay in the hospital (Straaten, Spoelstra-de Man, and de Waard 2014).

Sepsis

Sepsis arises due to complications of an infection, such as bacterial growth, which can become life threatening. Macrophages are key cells in the immune system that engulf and destroy target cells that may cause infection. They also produce ROS, which are usually necessary to control infections. If left uncontrolled, such infections are the leading cause of sepsis. It was discovered that incubation of macrophages with ascorbate helps to regulate the macrophage process significantly. Furthermore, ascorbate exhibited strong bacteriostatic activity, which further aids the body in fighting infections (Straaten, Spoelstra-de Man, and de Waard 2014).

Overall Well-Being and Physical Activity

Vitamin C has been noted to play a role in promoting physical activity, which may relate to the fact that fatigue and oxidative stress go hand in hand. Therefore, antioxidant-rich vitamin C helps to quell some fatigue. Additionally, vitamin C boosts the brain's oxidative fuel supply, since vitamin C characteristically contains neuroprotective properties that may promote a sense of well-being (Johnston, Barkyoumb, and Schumacher 2014).

The Possibility: Brain Protection with Aging

Vitamin C initiates the immediate antioxidant response to oxidative signaling. Before free radicals are able to attack lipids, vitamin C inhibits their course of action through its aqueous phase. Its antioxidant features protect cells from damage caused by oxidative stress through the neutralization of lipid hydroperoxyl radicals and by protecting proteins from other forms of damaging peroxidation reactions.

There are multiple pathological procedures that may induce the production of ROS in the brain. These include but are not limited to depletions of antioxidant supply and vascular degenerative disease or cognitive complications correlated with old age activated by oxidative and nitrosative stress. Dementia and Alzheimer's disease are perhaps the most striking examples for these symptoms. Dementia is described as a general decline in cognitive function severe enough to disrupt the flow of daily activities. Though the two disorders are similar, dementia is a cataloged group of symptoms, whereas Alzheimer's is a progressive disease that slowly causes cognitive degeneration. It begins with memory loss and then eventually disturbs other mental faculties. Amyloids are proteins found regularly dispensed throughout the body. Unfortunately, when they divide improperly, forming their beta structure, the buildup accumulates in extracellular matter, causing toxic damage to brain neurons. Intraneuronal neurofibrillary tangles occur when the protein undergoes overphosphorylation and aggregates into an insoluble group. These, along with the loss of synaptic function, are the primary markers for Alzheimer's disease. In the very beginning stages of Alzheimer's, oxidative stress is notable. As oxidative stress develops the effects produce mild expressions of cognitive impairment. This supports the claim that preventative measures addressing oxidative stress should be a priority in public health. Safe and effective methods to reduce oxidative damage early on are a necessary part of combating neurodegenerative disorders.

The concentration of vitamin C in the brain is extremely high and difficult to deplete; the only organ with comparable amounts is the adrenal gland. Studies involving animals have shown that the brain retains vitamin C even at the expense of other organs. Under specific conditions of possible depletion, certain areas, such as the hippocampus, cerebellum, and cortex, which are usually targeted in disease environments, retain vitamin C most effectively. However, if there happen to be chronic levels of instability or stress, the brain is no longer able to maintain appropriate levels of vitamin C. These cases may result in pathological aging and neurodegenerative disorders. Both human and animal studies demonstrate that ascorbic acid deficiency is associated with oxidative stress markers, and oxidative stress is a consistent observation in Alzheimer's disease (Harrison, Bowman, and Polidori 2014). This is not to suggest that deficiencies in vitamin C alone are a direct cause of Alzheimer's and other disorders of the brain but rather to highlight their relationship. For instance, studies have also shown that the oxidized form of vitamin C may have an effect on antioxidant levels in patients with cerebral ischemia. Dehydroascorbic acid, the oxidized form of ascorbic acid, has been shown to cross the blood-brain barrier by means of facilitative transport and may offer neuroprotection against cerebral ischemia by augmenting antioxidant levels in the brain (Naidu 2003).

Food Preservation and Other Uses

As we mentioned earlier, ascorbic acid has many other uses. For example, it works as a preservative and color stabilizer for various foods and beverages. Ascorbic acid has the ability to target enzymes that degrade fruits and vegetables. For example, it can prevent the surface of an apple from turning brown after being cut. Since ascorbic acid has the ability to neutralize oxygen, when used as a preservative, it helps to slow or stop the ripening process, which is caused by the presence of oxygen. Oxygen is also vital for many microorganisms to thrive, so ascorbic acid delays the decaying process.

Ascorbic acid is also used to preserve the red, fleshy color of meat, as it blocks the meat's propensity to develop carcinogens. The additive also helps to preserve the fresh taste of the meat. Manufacturers also add ascorbic acid to many items we consume daily, such as canned vegetables, jams, bottled juices, and other preserved fruits, to preserve freshness. Ascorbic acid prevents the phenolase enzyme from being activated. If this enzyme were activated, it would rapidly accelerate the oxidation process of the stored item. For many of us, our intake of vitamin C is highly reliant on these additives.

Daily Requirements and Toxicity

In 1979, the Food and Drug Administration's recommended daily allowance (RDA) for vitamin C was 45 mg for both men and women (Cameron, Pauling, and Leibovitz 1979). However, as the years progressed, this figure has gradually increased. It is now 90 mg for men over the age of nineteen and 75 mg for women. However, it has been recommended that smokers consume at least 140mg/day of ascorbic acid in order to maintain a total body pool similar to that of the nonsmoker consuming the recommended amount. Stress, alcoholism, fever, and viral infections are known to cause a rapid decline in blood levels of ascorbic acid (Naidu 2003; Kallner et al. 1981).

Some researchers believe that in order to reduce the risk of noncommunicable diseases and maintain adequate saturation of the cells, the recommended daily intake of vitamin C should be increased to 100 to 120 mg. However, we can base our required intake on our current health status and lifestyle. Fortunately, vitamin C is nontoxic. However, at high doses, such as doses of more than 2 g per day, diarrhea and gastrointestinal disturbances can manifest. These side effects are generally not serious and can be alleviated by reducing one's intake. Thus far, there is no concrete evidence that vitamin C has serious negative health effects (Chambial et al. 2013).

2013 RDAs by Age Based on Most Recent Recommendations

Age	Male	Female	Pregnant	Lactating
0–6 months	40 mg*	40 mg*		
7–12 months	50 mg*	50 mg*		
1–3 years	15 mg	15 mg		
4–8 years	25 mg	25 mg		
9–13 years	45 mg	45 mg		
14–18 years	75 mg	65 mg	80 mg	115 mg
19+ years	90 mg	75 mg	85 mg	120 mg
Smokers	Individuals who smoke require 35 mg per day more vitamin C than nonsmokers.			

* Adequate intake (AI). Table taken from the National Institutes of Health Office of Dietary Supplements (https://ods.od.nih.gov/factsheets/VitaminC-HealthProfessional/).

Food Prep

We have a tendency to overcook our vegetables; this tendency can reduce the amount of nutrients available for our bodies to absorb. For both vegetables and fruits, vitamin content depends on the type of fruit or vegetable, its maturity at harvest, genetic variation, growing conditions, postharvest handling, storage conditions, processing, and preparation (Howard et al. 2008). Vitamin C content is also influenced by pH and the presence of transition metals, such as copper. The ideal pH for vitamin C is between four and six (Naidu 2003).

Vitamin C is denatured easily; therefore, excessive amounts of heat can destroy any traces of the vitamin completely. Sunlight, high temperatures, and oxygen from the air all cause vitamin C to become oxidized. For example, when boiling vegetables, the vitamin C easily leaches into the water, since vitamin C is highly soluble in water. Many of us cook based on our taste and convenience; however, in order to retain high quantities of vitamin C, it is recommended that you cook your vegetables over a low heat using small amounts of water for short periods to maximize your vitamin C content (Igwemmar, Kolawole, and Imran 2013).

Vitamin C's potential has been well documented for centuries. It is arguably one of the most affordable and easily accessible vitamins available. Its role within our bodies facilitates numerous processes ranging from potentiating enzyme reactions to facilitating immunity defense mechanisms. It is a soldier in the body's army, scavenging free radicals to protect the body's cellular structures from damage that arises from infections and aging. It is a vitamin that is in high demand, and due to its inability to be stored in the body, it must be replaced on a daily basis through the consumption of vitamin C–rich fruits and vegetables or proper supplementation. As we age, our bodies retain less vitamin C, and therefore we should increase intake to protect us from many of the noncommunicable diseases that plague our aging process. Furthermore, smokers should pay close attention to their intake, as their bodies are starved of vitamin C and need more support for their cardiovascular systems. This should not be a challenge for many, as vegetables like peppers and fruits like guava are extremely rich sources of vitamin C. The more popular fruits, such as apples, pears, grapes, and bananas, contain much less vitamin C (Szeto et al. 2002).

Finally, when purchasing imported fruits and vegetables, it is important to be mindful that the amount of ascorbic acid may be lower than it should be due to poor transportation and storage facilities. Vitamin C is the third and final piece of our life-pill puzzle; its addition only reinforces the immense potential of this capsule.

Perfection is rarely attained, but a balance of forces may suffice.

—Alfred Sparman, MD

10

The Life Pill

As doctors, we have a tendency to focus on patching up the illness at hand while neglecting the etiology of the disease process. This leads to the vicious cycle of recurrent disease and the propagation of many preventable illnesses. Throughout the book, we have used scientific data and research that emphasized the power of antioxidants. Oxidation appears to be the initial culprit in most diseases known to humanity. This occurs due to oxidative stress and the production of damaging free radicals. However, in a balanced state, the formation of free radicals in the presence of sufficient antioxidants should ensure the perfect environment in which to repair or replace cells. In this manner, our body's mechanisms will function optimally.

Let's recap electrons. Electrons in the outermost shell of an atom like to be paired. If they are unpaired, the atom becomes unstable and seek an extra electron to make that outer shell stable. This process of unstable atoms and molecules produces oxidative stress, which can also occur in very stressful environments or exposure to UV light, pollution, pesticides, lead, mercury, contaminated water, and so forth. Oxygen seems to be one of the main elements in free radical formation; however, hydrogen and nitrogen or RNS can also be involved. Once free radicals are formed, they behave as scavengers, trying to steal electrons from stable cells, and this domino effect continues. It can only be halted if the reaction comes into contact with an antioxidant. I should mention that in the absence of antioxidants, the cell tries to repair itself, which in many instances affects, producing the defective coding that results in diseases, including cancer and diabetes. The body has a secret weapon: one of the most important antioxidants is SOD. Others include glutathione, catalase, and ALA. However, if these antioxidants are overwhelmed, the proliferation of free radicals and cell instability ensue.

When a free radical interacts with an antioxidant, the antioxidant gives it an electron, causing the free radical to be reduced and the antioxidant to be oxidized. This is called a redox reaction. It is well-known that the oxidation of LDL particles produces atherosclerosis, which leads to heart disease. Additionally, when proteins and fats are oxidized, wrinkling of the skin results, and DNA oxidation leads to cancer formation. So even though we need oxidation within the body, the substrate that is oxidized must be in an extremely controlled environment to avoid excessive free radical production.

People who eat a Western diet with high amounts of trans fats, processed foods, and saturated fats have shorter life spans due to many noncommunicable diseases. However, people who eat more plant product, less red meat, more fish, and more monounsaturated fats and polyunsaturated fats, such as most in Japan, tend to have longer life spans. It is clear that how healthily and how long we live depends on many factors, but one of the major factors appears to be the quality and type of food we eat.

The life pill is formulated to naturally combat these diseases with a combination of three super agents. The first is vitamin C.

Vitamin C

Vitamin C is one of the most potent antioxidants in the world. Therefore, it is easy to see why its daily intake is beneficial. Its uses include the following:

- collagen synthesis and the integrity of cell walls
- assisting with the absorption of iron in the gut to form blood cells
- decreasing cardiovascular disease
- helping to improve endothelial functions
- diabetes control
- shortening the length and severity of the common cold
- boosting the immune system
- helping with anemia
- preventing neurological disorders
- ameliorating kidney disease
- preventing and treating Alzheimer's disease and dementia
- enhancing men's fertility
- promoting health during pregnancy
- avoiding sepsis
- aiding chronic smokers
- overall well-being and food preservation
- helping prevent cancer and heart disease
- detoxifying the body
- supporting the good bacteria in your gut
- killing candida, bacteria, fungi, viruses, and parasites
- preventing the hardening of arteries
- neutralizing harmful environmental and toxins
- the destruction of free radicals
- combating stress
- acting as an antidepressant
- removing heavy metals like mercury and lead
- lowering high cholesterol

Moringa oleifera

There has recently been tremendous buzz over the use of *Moringa*, which is a powerful plant. It has forty-six types of antioxidants, including high doses of vitamin C and vitamin A. There are eighteen amino acids present, eight of which are essential amino acids, making it a complete protein; this is a rarity in the plant world! Indeed,

Moringa's protein content rivals that of meat, making it an excellent source of essential protein for vegetarians and vegans.

Moringa is a source of the following nutrients:

- carbohydrates
- proteins
- fats
- arginine
- histidine
- isoleucine
- leucine
- lysine
- methionine
- phenylalanine
- threonine
- tryptophan
- valine
- carotene (vitamin A)
- vitamin C
- thiamine (vitamin B1)
- riboflavin (vitamin B2)
- niacin (vitamin B3)
- copper
- calcium
- fiber
- iron
- magnesium
- phosphorous
- potassium
- zinc

Initial research shows that it aids with the following processes:

- diabetes and glucose control
- asthma treatment

- antifungal treatment
- antibacterial activity
- antimicrobial activity
- parasite detoxification
- cancer and tumor treatment and prevention
- blood pressure control
- cholesterol control
- liver protection
- cardiac disease treatment and prevention
- thyroid dysfunction treatment
- diuretic activity
- ulcer and gastritis treatment
- inflammation treatment
- postnatal milk production
- anemia treatment
- energy production
- rheumatism treatment
- joint pain relief
- edema treatment
- arthritis treatment
- improving the immune system
- weight loss
- skin nourishment
- enhancement of male sexual function

Here again is a superior antioxidant. In recent years, the *Moringa oleifera* craze has engulfed the populace. Men and women are using *Moringa oleifera* to alleviate illnesses, especially hypertension, diabetes, and the common cold, and to enhance sexual function. However, they have had no idea of the daily amount necessary to optimize their reward from this great plant. Their questions are now answered.

Bryophyllum pinnatum

This unique plant contains newly discovered derivatives that have the potential for future medicinal use. It has been around for centuries; however, only recently has it been made available on the market. This plant also is called the life plant, the wonder of the world, the good luck plant, the resurrection plant, and Africa never dies, because it has the ability to regenerate in extremely hostile environments when other plants would perish. Without water, it produces green leaves. Once, I placed a leaf of this plant on my patio on a zinc sheet under the blazing sun; new roots began to grow without access to water. My seven-year-old daughter said, "Dad, that plant will live forever!" I say this to trigger your imagination. If a plant can reproduce itself in such a fashion, then it must have properties that can stimulate cells to reproduce.

Its cardiac glycoside composition would contribute to its use for weak hearts. Glycosides in other tablet forms are currently used for heart failure conventionally. Its antioxidant properties stabilize endothelial cells in and around the heart by causing the release of nitrous oxide, the most potent vasodilator in the body. This results in increased arterial dilatation and increased heart flow. People use it without even knowing its properties; however, one thing they know is that it heals their diseases. If a plant can regenerate like that without intervention, if a plant can live forever like that, then its nutrients must contain the machinery for humans *almost* to do the same. This is the first time that science has introduced this life-giving plant to the people.

Bryophyllum pinnatum is composed of the following substances:

- carbohydrates
- crude fiber
- proteins
- fat
- substantial amounts of minerals, vitamins, and amino acids

- cardiac glycosides
- sodium
- calcium
- potassium
- phosphorous
- magnesium
- manganese
- iron
- copper
- zinc

The vitamins and phytochemicals include the following:

- ascorbic acid
- riboflavin
- thiamin
- niacin
- alkaloids
- flavonoids
- phenols
- tannins

It is used as a treatment for the following ailments:

- hypertension
- high cholesterol
- diabetes
- wounds
- pain
- bladder stones
- gonorrhea
- rheumatoid arthritis
- stomach bugs
- the common cold
- diarrhea
- vomiting

- gastric ulcers
- asthma
- respiratory infections
- palpitations
- neurological disorders and muscle tension
- depression
- headaches
- edema of the legs
- cancer
- gallstones
- anemia
- eczema
- bronchitis
- bacterial, fungal, and viral infections
- kidney stones
- skin problems
- hemorrhoids
- ailing hearts
- sexual dysfunction

The life pill—as part of a healthy lifestyle—can lead to a better quality of life and the possibility of a longer life span. If your desire is a long, healthy life, you must turn things around *now*. Let's look at how the life pill, along with lifestyle changes, can be of benefit to you.

Interestingly, the combination of seven colors gives you white light. The combination of seven lifestyle changes and the life pill ensures longevity.

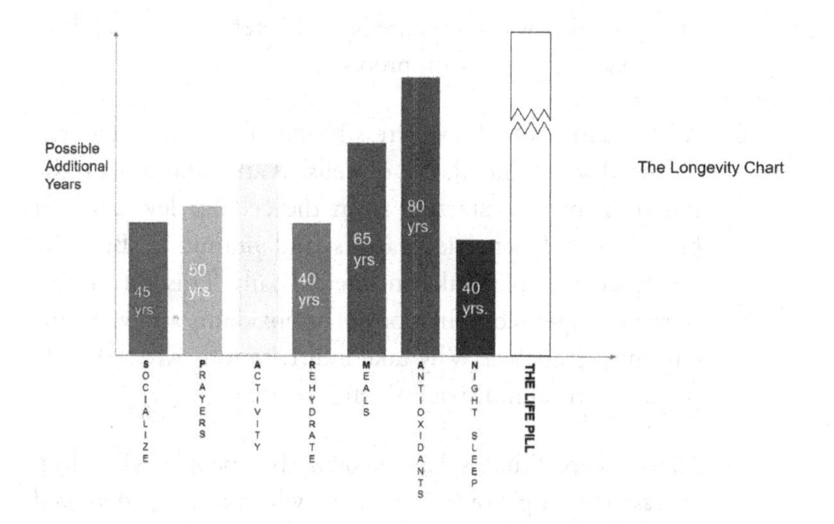

There are seven steps to longevity. Here they are:

1. **S**ocialize. Recall that people who are single live shorter lives than married people. Having a good support system diffuses stress. So call a friend, speak to your siblings and close relatives, dance, laugh, sing, and listen to music.

2. **P**ray or meditate. People who pray and allow a higher power to guide their lives are less burdened, as they know that someone bigger is in control to handle their cares.

3. **A**ctivity. The American Heart Association recommends twenty to thirty minutes of brisk exercise at least three times weekly. So climb the stairs, walk the dog, and get involved in exercises that increase your heart rate.

4. **R**ehydrate. Drink four to six glasses of water daily. This will help to detoxify your body and maintain fluid balance.

5. **M**eals. Consumption of red meat is associated with almost all cancers, not to mention cardiovascular disease. Our teeth share a common morphology with those of herbivores, so

maybe tearing was not in the plan. Therefore, we should eat less meat and more plant products.

6. **A**ntioxidants. Oxidative stress is one of the major culprits in the disease and death of cells. Antioxidants halt the mutation process starting from the cellular level. So eat blackberries, blueberries, carrots, and vitamin C. But why not make it simple? Take one life pill daily. This pill has one of the strongest combinations of antioxidants and vitamins currently available. Why take multivitamins when the life pill is a natural multivitamin and more?

7. **N**ight sleep. Studies have shown that people who sleep excessively long hours and those who are sleep deprived have shortened life spans. So get seven hours of sleep each night—no more and no less.

These seven steps must be followed constantly. This is the lifestyle that will lead to longer living. The life pill is the most important part of this entire formula; by saying this I mean that antioxidants taken on a regular basis will decrease your risk of disease and promote longevity. These three medicinal plant products will have a synergistic effect in combating noncommunicable diseases, such as cancer, diabetes, hypertension, high cholesterol, erectile dysfunction, arthritis, and others that I have previously mentioned. The daily use of this product in combination with lifestyle modifications will not only enhance your health but also will offer you a longer life without any side effects. It will give and not take; it will reduce spending on unnecessary pills. As a cardiologist who practices conventional medicine, I highly recommend "The Life Pill"! Whether you have diseases or not, the pill will work for you as a preventative measure or as an aid in combating your ailments. It is what was intended for you in the first place. It is what the world is waiting for. It is "The Life Pill"!

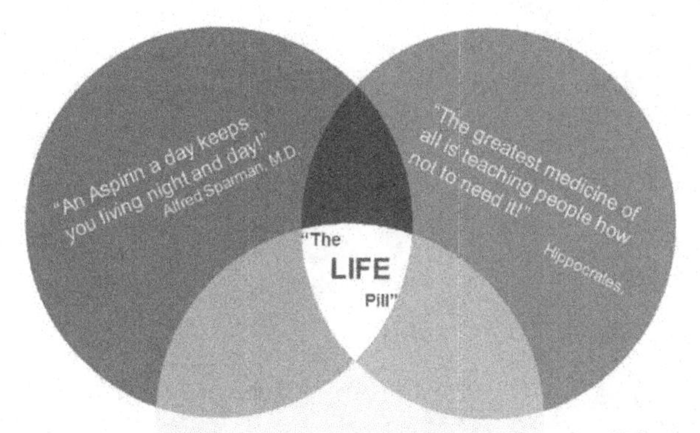
"An Aspirin a day keeps
you living night and day!"
Alfred Sparman, M.D.
"The greatest medicine of
all is teaching people how
not to need it!"
Hippocrates.
"The
LIFE
Pill"
"Keep Close to the Plant."
Alfred Sparman, M.D.

"Why Not Take Life for Life?"
THE LONGEVITY CHART

Glossary

abortifacient: A substance (typically a drug) that causes abortions.

adenosine diphosphate (ADP): An important organic compound in metabolism essential to the flow of energy in living cells.

adenosine triphosphate (ATP): A nucleoside triphosphate used in cells as a coenzyme, considered the molecular unit of currency of intracellular energy transfer. ATP transports chemical energy within cells for metabolism.

adsorption: The binding of molecules or particles to the surface of solid bodies or liquids with which they are in contact.

aerobic: Relating to, involving, or requiring free oxygen.

alley cropping: The planting of rows of trees widely spaced with a companion crop grown in between the rows.

alpha-keto acids: Intermediate products formed during the process of cells converting amino acids into energy.

alpha-lipoic acid (ALA): A compound primarily made in the body and found in every cell, it helps turn glucose into energy. It is classed as an antioxidant.

amino acids: The building blocks of proteins.

anabolism: The set of metabolic pathways that construct molecules from smaller units. These reactions require energy.

anaerobic: Without oxygen.

angular momentum: The quantity of rotation of a body, which is the product of its moment of inertia and its angular velocity.

antinociceptive: Reducing sensitivity to painful stimuli.

antioxidant: A substance, such as vitamin C or E, that removes potentially damaging oxidizing agents in a living organism.

apoptosis: The death of cells that occurs as a normal and controlled part of an organism's growth or development.

atherogenesis: The process of forming atheromas, plaques in the inner lining (the intima) of arteries.

atomic mass: The mass of an atom of a chemical element expressed in atomic mass units. It is approximately equivalent to the number of protons and neutrons in the atom (the mass number), or to the average number allowing for the relative abundances of different isotopes.

atomic orbitals: The probability distribution of an electron in an atom or molecule.

atomic radius: A measure of the size of an element's atoms, usually the mean or typical distance from the center of the nucleus to the boundary of the surrounding cloud of electrons.

ATP synthase: An enzyme that catalyzes the formation of ATP from the phosphorylation of ADP with inorganic phosphate using a form of energy, such as the energy from a proton gradient.

atrophy: To waste away, typically due to the degeneration of cells, or to become vestigial during evolution.

autocrine signaling: A form of cell signaling in which a cell secretes a hormone or chemical messenger (called the autocrine agent) that binds to autocrine receptors on that same cell, leading to changes in the cell.

autooxidation: Spontaneous oxidation of a substance at ambient temperatures in the presence of oxygen.

ayuverdic medicine: One of the world's oldest holistic healing systems. It was developed thousands of years ago in India. It is based on the belief that health and wellness depend on a delicate balance between the mind, body, and spirit. Also called ayurveda.

binding energy: The energy that holds a nucleus together. Equal to the mass defect of the nucleus.

biogas: A mixture of different gases produced by the breakdown of organic matter in the absence of oxygen. It can be produced from raw materials, such as agricultural waste, manure, municipal waste, plant material, sewage, green waste, or food waste.

bisexual: Having both male and female structures or combining both sexes in one structure. Flowers of this kind are referred to as perfect, having both stamens and carpels.

Bitot's spots: Shiny, pearly spots of triangular shape occurring on the conjunctiva in severe vitamin A deficiency, especially in children.

blebbing: The process in which a cell detaches its cytoskeleton from the membrane, causing the membrane to swell into spherical bubbles, greatly distorting the shape of the cell. It usually occurs when a cell is about to undergo apoptosis.

calcium: A silver-white bivalent metallic element that is an alkaline earth metal, occurs only in combination, and is an essential constituent of most plants and animals.

carbohydrates: One of the main types of nutrients and the most important source of energy for the body. The digestive system changes carbohydrates into glucose, which it uses for energy for cells, tissues, and organs.

carbon: A naturally abundant, nonmetallic element that occurs in all organic compounds and can be found in all known forms of life.

carcinogenesis: The initiation of cancer formation.

carotenoid: Any of a class of mainly yellow, orange, or red fat-soluble pigments, including carotene, which give color to plant parts, such as ripe tomatoes and autumn leaves.

carrageenan: A family of linear sulfate polysaccharides extracted from red edible seaweeds. They are widely used in the food industry for their gelling, thickening, and stabilizing properties.

catabolism: The set of metabolic pathways that breaks down molecules into smaller units that are either oxidized to release energy or used in other anabolic reactions.

catalase: A common enzyme found in all living things that catalyzes the decomposition of hydrogen peroxide to water and oxygen.

cell: The basic structural and functional unit of an organism, typically microscopic and consisting of cytoplasm and a nucleus enclosed in a membrane.

cellular necrosis: A type of cell death that lacks the features of apoptosis and autophagy and is usually considered to be uncontrolled.

cellular respiration: What cells do to break up sugars into a form that the cells can use as energy. This happens in all forms of life. Cellular respiration takes in food and uses it to create ATP, a chemical that cells use for energy.

chelating antioxidants: Antioxidants that work by decreasing the prooxidant effect by reducing the redox potential and stabilizing the oxidized form of a metal.

chemical compound: A substance consisting of two or more different chemically bonded chemical elements, with a fixed ratio determining the composition.

chromosomes: Threadlike structures located inside the nucleus of animal and plant cells. Each chromosome is made of protein and a single molecule of DNA. Passed from parents to offspring, DNA contains the specific instructions that make each type of living creature unique.

chronological: Arranged in order of time.

coagulants: Substances that cause blood or another liquid to change to a solid or semisolid state.

coenzyme Q10 (CoQ10): An antioxidant cells use to produce the energy your body needs for cell growth and maintenance. Also known as ubiquinone.

coenzyme: A substance that enhances the action of an enzyme.

collagen: The most abundant protein in the human body and the substance that holds the whole body together. It is found in the bones, muscles, skin, and tendons, where it forms a scaffold to provide strength and structure.

conglomerate rock: A sedimentary clastic rock consisting of rounded fragments of preexisting rocks larger than 4-mm set within a more finely grained sedimentary matrix.

connective tissue: Tissue that connects, supports, binds, or separates other tissues or organs, typically having relatively few cells embedded in an amorphous matrix, often with collagen or other fibers, and including cartilaginous, fatty, and elastic tissues.

copper/zinc superoxide dismutase (CuZnSOD): An oxidoreductase enzyme responsible for the very rapid two-step dismutation of the toxic superoxide radical to molecular oxygen and hydrogen peroxide through alternate reduction and oxidation of the active-site copper.

covalent bond: A bond involving the sharing of a pair of valence electrons by two atoms.

culturing: The cultivation of microorganisms, such as bacteria, or of tissues for scientific study, medicinal use, and so forth.

cysteine: A sulfur-containing amino acid that occurs in keratins and other proteins, often in the form of cystine, and is a constituent of many enzymes.

cytokinesis: The cytoplasmic division of a cell at the end of mitosis or meiosis, bringing about its separation into two daughter cells.

cytoplasm: A thick solution that fills each cell and is enclosed by the cell membrane. It is mainly composed of water, salts, and proteins.

de Broglie wave: In 1924, a young physicist, de Broglie, speculated that nature did not single out light as the only matter that exhibits a wave-particle duality. He proposed that ordinary particles, such as electrons, protons, or bowling balls, could also exhibit wave characteristics in certain circumstances.

deamination: The process by which amino acids are broken down if there is excess protein intake. The amino group is removed from the amino acid and converted to ammonia. The rest of the amino acid is made up of mostly carbon and hydrogen and is recycled or oxidized for energy.

delta: A usually triangular mass of sediment, especially silt and sand, deposited at the mouth of a river.

diabetic ketoacidosis: A serious diabetes complication in which the body produces excess blood acids (ketones).

dismutation: A process of simultaneous oxidation and reduction—used especially of compounds taking part in biological processes.

DNA: Deoxyribonucleic acid is a molecule that carries most of the genetic instructions used in the development, functioning, and reproduction of all known living organisms and many viruses.

duodenum: The first part of the small intestine immediately beyond the stomach, leading to the jejunum.

edema: Swelling caused by excess fluid trapped in your body's tissues. Although edema can affect any part of your body, it's most commonly noticed in the hands, arms, feet, ankles, and legs.

electronegativity: A measure of the tendency of an atom to attract a bonding pair of electrons.

electron carriers: Any of various molecules that are capable of accepting one or two electrons from one molecule and donating them to another in the process of electron transport. As the electrons are transferred from one electron carrier to another, their energy level decreases, and energy is released.

electron cloud: An informal term in physics used to describe where electrons are when they go around the nucleus of an atom.

electron configuration: In atomic physics and quantum chemistry, the distribution of electrons of an atom or molecule (or other physical structure) in atomic or molecular orbitals.

electron: A negatively charged elementary particle.

embryo: An unborn or unhatched offspring in the process of development.

endocrine signaling: Endocrine signaling occurs when endocrine cells release hormones that act on distant target cells in the body.

enzymatic antioxidants: Antioxidants produced naturally by the body.

enzymes: Biological molecules (proteins) that act as catalysts and help complex reactions occur everywhere in life.

epithelial tissue: Tissue that covers the whole surface of the body. It is made up of cells closely packed and ranged in one or more layers. This tissue is specialized to form the covering or lining of all internal and external body surfaces.

fallopian tube: In a female mammal, either of a pair of tubes along which eggs travel from the ovaries to the uterus.

fecundity: The ability to produce abundant, healthy growth or offspring.

Fenton reaction: H. J. H. Fenton discovered in 1894 that several metals have a special oxygen-transfer properties that improve the use of hydrogen peroxide. Actually, some metals have a strong catalytic power to generate highly reactive hydroxyl radicals. Since this discovery, the iron catalyzed hydrogen peroxide has been called Fenton's reagent.

fibroblasts: A cell in connective tissue that produces collagen and other fibers.

flocculate: To form or cause to form into small clumps or masses.

foliar nutrient: A nutrient supplied to plants through their foliage. Typically, it involves spraying water-dissolved fertilizers directly onto the leaves.

free radical theory of aging (FRTA): The theory that organisms age because cells accumulate free radical damage over time.

free radical: An uncharged molecule (typically highly reactive and short-lived) with an unpaired valence electron.

genes: The basic physical and functional units of heredity. Genes are made up of DNA and act as instructions to make molecules called proteins. In humans, genes vary in size from a few hundred DNA bases to more than 2 million bases.

genome: An organism's complete set of DNA, including all its genes. Each genome contains all the information needed to build and maintain that organism.

genotype: The genetic makeup of an organism or group of organisms with reference to a single trait, set of traits, or entire complex of traits.

gerontology: The study of the social, psychological, cognitive, and biological aspects of aging.

gliadin: A class of proteins present in wheat and several other cereals within the grass genus *Triticum.*

gluconeogenesis: The synthesis of glucose from noncarbohydrate sources, such as amino acids and glycerol. It occurs primarily in the liver and kidneys whenever the supply of carbohydrates is insufficient to meet the body's energy needs.

glutathione (GSH): An important antioxidant in plants, animals, fungi, and some bacteria and archaea that prevents damage to important cellular components caused by reactive oxygen species, such as free radicals, peroxides, lipid peroxides, and heavy metals.

glutathione reductase: An enzyme that plays a crucial role in protecting the body against oxidative damage.

hemodynamic: Of or relating to the flow of blood within the organs and tissues of the body.

hemolysis: The rupture or destruction of red blood cells.

hepatoprotection: Prevention of damage to the liver. This damage is known as hepatotoxicity.

high-density lipoproteins (HDL): Cholesterol that absorbs other cholesterol and carries it back to the liver, which flushes it from the body. HDL is known as good cholesterol, because having high levels can reduce the risk for heart disease and stroke.

homolytic bond cleavage: Bond breaking in which the bonding electron pair is split evenly between the products. Homolytic cleavage often produces radicals.

homonuclear: Composed of only one type of element.

Hund's rule: An observational rule that states that a greater total spin state usually makes the resulting atom more stable. Accordingly, if two or more orbitals of equal energy are available, electrons will occupy them singly before filling them in pairs.

hydrogen: The simplest and lightest of the elements. It is normally a colorless, odorless, highly flammable diatomic gas.

hydrogen bond: A weak bond between two molecules resulting from an electrostatic attraction between a proton in one molecule and an electronegative atom in the other.

hydrophilic: Having a tendency to mix with, dissolve in, or be wetted by water.

hydrophobic: Tending to repel or fail to mix with water.

hyperoxia: Exposure to an excess supply of oxygen or higher than normal partial pressure of oxygen.

hyperthyroidism: The overproduction of a hormone by the thyroid.

hypothyroidism: A condition in which the thyroid gland doesn't produce enough thyroid hormone.

in vitro: Performed or taking place in a test tube, culture dish, or elsewhere outside a living organism.

in vivo: Performed or taking place in a living organism.

infundibulum: The hollow stalk that connects the hypothalamus and the posterior pituitary gland.

inorganic: Not consisting of or derived from living matter.

inorganic compound: Any substance in which two or more chemical elements other than carbon are combined, nearly always in definite proportions.

interphase: The resting phase between successive mitotic divisions of a cell or between the first and second divisions of meiosis.

interstitial cells of Leydig: Endocrine cells that mainly produce testosterone, releasing it into the blood and neighboring tissues.

intracranial hemorrhage: Rupture of a blood vessel within the brain or between the skull and the brain.

ionic bond: The electrostatic bond between two ions formed through the transfer of one or more electrons.

isotopes: Atoms that have the same number of protons but different numbers of neutrons.

ketone bodies: An intermediate product of fatty acid metabolism. Ketone bodies tend to accumulate in the blood and urine of individuals affected by starvation or uncontrolled diabetes mellitus.

law of multiple proportions: A basic law of chemistry that states when two elements combine to form more than one compound, the ratios of the masses of one element that combine with a fixed mass of the other will be ratios of small whole numbers.

Lewis theory of bonding: The theory that chemical bonds are formed when atoms transfer (ionic bonding) or share (covalent bonding) valence electrons to attain noble gas electron configurations.

lipids: Naturally occurring molecules that include fats; waxes; sterols; fat-soluble vitamins (such as vitamins A, D, E, and K); monoglycerides; diglycerides; triglycerides; phospholipids; and others. Lipids are insoluble in water.

lipoic acid: A naturally occurring antioxidant compound produced in the body and synthesized by both plants and animals. It is vital to cellular energy production and helps to neutralize the damage caused by free radicals.

lipophilic: Tending to combine with or dissolve in lipids or fats.

low-density lipoproteins (LDL): Known as bad cholesterol, because having high levels can lead to plaque buildup in your arteries and result in heart disease and stroke. Makes up the majority of cholesterol in the body.

lymphocyte: A form of small leukocyte (white blood cell) with a single, round nucleus that occurs especially in the lymphatic system.

magnetic moment: The property of a magnet that interacts with an applied field to give a mechanical moment.

manganese superoxide dismutase (MnSOD): The primary mitochondrial ROS-scavenging enzyme that converts superoxide to hydrogen peroxide, which is subsequently converted to water by catalase and other peroxidases.

melanin: The pigment that gives human skin, hair, and eyes their color. Dark-skinned people have more melanin in their skin than light-skinned people have. Melanin is produced by cells called melanocytes.

mevalonate pathway: The pathway by which isoprenoids are produced; these diverse compounds include over thirty thousand biomolecules, including cholesterol, vitamin K, coenzyme Q10, and all steroid hormones.

metabolism: The chemical processes that go on continuously inside the body to keep you alive and your organs functioning normally, such as breathing, repairing cells, and digesting food.

metallic bond: A type of chemical bond that occurs between atoms of metallic elements. It gives metals their unique properties, which we do not see in nonmetal substances.

mineral: A chemical substance (such as iron or zinc) that occurs naturally in certain foods and that is important for good health.

mitochondrion: An organelle found in large numbers in most cells, in which the biochemical processes of respiration and energy production occur.

mitogens: Substances that induce or stimulate mitosis.

mitosis: A type of cell division that results in two daughter cells each having the same number and kind of chromosomes as the parent nucleus, typical of ordinary tissue growth.

monosaccharides: The simplest form of carbohydrates. They consist of one sugar and are usually colorless, water-soluble, crystalline solids. Examples are glucose and fructose.

morula: An early-stage embryo consisting of cells (called blastomeres) in a solid ball.

multiple-function antioxidants: Substances that have more than one mechanism of antioxidant activity.

murine animals: Members of the rodent subfamily *Murinae,* which includes the house mouse and the brown rat.

muscle tissue: Tissue composed of cells or fibers that contract to produce movement in the body.

mutation: In biology, a permanent change of the genome of an organism, virus, or of extrachromosomal DNA or other genetic elements.

nerve tissue: Highly differentiated tissue composed of nerve cells, nerve fibers, dendrites, and neuroglia.

network covalent bonding: The formation by covalent bonds of a continuous network extending throughout the material. In a network solid, there are no individual molecules, and the entire crystal may be considered a macromolecule.

neutron: An elementary particle without a charge.

niazimicin: A compound found within the seeds of *Moringa oleifera.*

nitrogen: The simplest and lightest of the elements, normally a colorless, odorless, diatomic gas.

nociceptive: Of, relating to, or denoting pain arising from the stimulation of nerve cells, often arising from damage or disease in the nerves themselves.

noncommunicable disease (NCD): A medical condition or disease that is noninfectious or nontransmittable. NCDs can be chronic diseases that last for long periods of time and progress slowly.

nucleobases: Nitrogen-containing biological compounds (nitrogenous bases) found linked to a sugar within nucleosides—the basic building blocks of deoxyribonucleic acid (DNA) and ribonucleic acid (RNA).

nucleophile: A chemical species that donates an electron pair to an electrophile to form a chemical bond in relation to a reaction.

nucleotide: One of the structural components, or building blocks, of DNA and RNA, it consists of a base (one of four chemicals—adenine, thymine, guanine, or cytosine), plus a molecule of sugar and one of phosphoric acid.

nucleus: In physics, the positively charged central core of an atom consisting of protons and neutrons and containing nearly all its mass. In biology, a dense organelle present in most eukaryotic cells, typically a single rounded structure bounded by a double membrane and containing the genetic material.

orbital angular momentum: The angular momentum of an electron in an atom or a nucleon in a nucleus that arises from its orbital motion rather than from its spin.

organ system: A group of organs that work together to perform one or more functions.

organelles: Any of a number of organized or specialized structures within a living cell.

organic: Of, relating to, or derived from living matter.

organic compound: Any member of a large class of gaseous, liquid, or solid chemical compounds whose molecules contain carbon.

organ: A part of an organism that is typically self-contained and has a specific vital function, such as the heart or liver in humans.

oxaloacetate: A dicarboxylic acid ketone that is an important metabolic intermediate of the citric acid cycle.

oxidants: An oxidant is a reactant that removes electrons from other reactants during a redox reaction.

oxidative phosphorylation: The metabolic pathway in which the mitochondria in cells use their structure, enzymes, and energy released by the oxidation of nutrients to reform ATP.

oxidative state: A number assigned to an element in a compound according to a number of rules. This number enables us to describe oxidation-reduction reactions and to balance redox chemical reactions.

oxidative stress: An imbalance between the production of free radicals and the ability of the body to counteract or detoxify their harmful effects through neutralization by antioxidants.

oxygen: A colorless, odorless, reactive gas and the life-supporting component of the air.

paracrine signaling: A form of cell-to-cell communication in which a cell produces a signal to induce changes in nearby cells, altering the behavior or differentiation of those cells.

Pauli's exclusion principle: The idea that no two electrons in an atom can be at the same time in the same state or configuration, proposed by the Austrian physicist Wolfgang Pauli to account for the observed patterns of light emission from atoms.

pentobarbital: A barbiturate prescription drug used to relieve tension, anxiety, nervousness, and trouble sleeping; it also helps with relaxation before having a surgery or medical procedure.

pentylenetetrazol: A drug formerly used as a circulatory and respiratory stimulant.

peptic ulcer: A sore that develops on the lining of the esophagus, stomach, or small intestine.

periodic table: A table of the chemical elements arranged in order of atomic number, usually in rows, so that elements with similar atomic structure (and hence similar chemical properties) appear in vertical columns.

perioxidation: Any oxidation reaction, especially of an oxide, that produces a peroxide.

phagocytosis: The ingestion of a smaller cell or cell fragment, a microorganism, or foreign particles by means of the local infolding of a cell's membrane and the protrusion of its cytoplasm around the fold until the material has been surrounded and engulfed by closure of the membrane and formation of a vacuole. It is characteristic of amoebas and some types of white blood cells.

phosphorus: An element that exists in two major forms—white phosphorus and red phosphorus—but, due to its high reactivity, is never found as a free element on earth.

phosphorylation: The addition of a phosphate group to a protein or organic molecule, which turns many protein enzymes on and off, thereby altering their function and activity.

photosynthesis: A process used by plants and other organisms to convert light energy, normally from the sun, into chemical energy that can be later released to fuel the organisms' activities.

phytochemicals: Chemical compounds that occur naturally in plants; some are responsible for color and other sensory experiences, such as the deep purple of blueberries and the smell of garlic.

pinnate: Having parts arranged on opposite sides of an axis.

polar covalent bond: A type of chemical bond in which a pair of electrons is unequally shared between two atoms.

polymer: Any of various chemical compounds made of smaller, identical molecules (called monomers) linked together. Some polymers, like cellulose, occur naturally, while others, like nylon, are artificial.

polymerization: A process of reacting monomer molecules together to form polymer chains or three-dimensional networks. There are many forms of polymerization, and different systems exist to categorize them.

polysaccharide: A carbohydrate (e.g., starch or cellulose) containing more than three monosaccharide units per molecule.

potentiation: The interaction between two or more drugs or agents resulting in a pharmacologic response greater than the sum of the individual response of each drug or agent.

primary antioxidants: Antioxidants that function by the donation of an electron or hydrogen atom to a radical derivative.

prokaryote: A microscopic, single-celled organism that has neither a distinct nucleus with a membrane nor other specialized organelles. Prokaryotes include bacteria and cyanobacteria.

prooxidant: A substance that accelerates the oxidation of another substance.

proton: A positively charged elementary particle.

pulmonary oxygen toxicity: A condition in the lungs resulting from the harmful effects of breathing molecular oxygen at elevated partial pressures.

pyruvate: A naturally occurring substance in the human body. It controls a person's metabolic rate and is produced in the liver.

quark: One of two currently recognized groups of fundamental particles, which are subatomic, indivisible (at least as far as we know today) particles that represent the smallest known units of matter.

quiescent cells: Cells in the G0 or resting phase. The G0 phase is viewed as either an extended G1 phase in which the cell is neither dividing nor preparing to divide or a distinct quiescent stage that occurs outside of the cell cycle.

rancidity: Spoilage of a food in such a way that it becomes undesirable (and usually unsafe) for consumption.

reactive nitrogen species (RNS): A subset of free oxygen radicals and are highly reactive. They are formed from nitric oxide and superoxide.

reactive oxygen species (ROS): Chemically reactive molecules containing oxygen. Examples include oxygen ions and peroxides. ROS are formed as a natural by-product of the normal metabolism of oxygen.

redox: A process in which one substance or molecule is reduced and another oxidized—oxidation and reduction considered together as complementary processes.

redox cycling: Repetitively coupled reduction and oxidation reactions, often involving oxygen and reactive oxygen species.

redox potential: A measure of the tendency of a chemical species to acquire electrons and thereby be reduced. Reduction potential is measured in volts (V) or millivolts (mV).

redox signaling: The concept that electron-transfer processes play a key messenger role in biological systems.

reducing agent: A substance that tends to bring about reduction by being oxidized and losing electrons.

resonance: A method of describing the delocalized electrons in some molecules in which the bonding cannot be explicitly expressed by a single Lewis structure.

resveratrol: A polyphenol compound found in certain plants and in red wine that has antioxidant properties and has been investigated for possible anticarcinogenic effects.

retinol: A yellow compound found in green and yellow vegetables, egg yolk, and fish-liver oil. It is essential for growth and vision in dim light.

retrolental fibroplasia: Abnormal proliferation of fibrous tissue immediately behind the lens of the eye, leading to blindness. It affected many premature babies in the 1950s owing to the excessive administration of oxygen.

rubefacient: A substance for topical application that produces redness of the skin—by causing dilation of the capillaries and an increase in blood circulation, for example.

saponin: A class of chemical compounds found in particular abundance in various plant species; these compounds have a distinctive foaming characteristic.

scavenging activity: A scavenger in chemistry is a chemical substance added to a mixture in order to remove or deactivate impurities and unwanted reaction products, such as oxygen, to make sure that they will not cause any unfavorable reactions.

scientific method: A body of techniques for investigating phenomena, acquiring new knowledge, or correcting and integrating previous knowledge. To be termed scientific, a method of inquiry must be based on empirical or measurable evidence subject to specific principles of reasoning.

secondary antioxidant: A substance that initiates antioxidant activity by the removal of an oxidative catalyst and the consequent prevention of the initiation of oxidation.

senescence: The condition or process of deterioration with age.

sepals: One of the usually green leaflike structures composing the outermost part of a flower. Sepals often enclose and protect the bud; some sepals may remain after the formation of the fruit.

sorption: The term used for both absorption and adsorption.

sp hybridization: One of the two hybrid orbitals formed by orbital fusion of an s orbital and a p orbital.

sp2: A type of hybridization in which a carbon atom is attached to three groups, resulting in three covalent bonds.

sp3: A type of hybridization in which a carbon atom is attached to four groups, resulting in four covalent bonds.

spermatozoa: A mature male germ cell that is the specific output of the testes and fertilizes the mature ovum during sexual reproduction. It is microscopic in size and looks similar to a translucent tadpole, with a flat, elliptical head containing a spherical center section and a long tail by which it propels itself with a vigorous lashing movement.

spin: A property of an electron that is loosely related to its spin about an axis. Two electron spin states are allowed with values of $+\frac{1}{2}$ or $-\frac{1}{2}$.

strong force: The strongest force of the four fundamental forces, it is the attractive force that keeps the repelling protons together in a nucleus.

subatomic particles: In the physical sciences, subatomic particles are particles much smaller than atoms. There are two types of subatomic particles: elementary particles, which, according to current theories, are not made of other particles and composite particles, which are made of other particles.

substrate: In biochemistry, the substrate is a molecule upon which an enzyme acts.

superoxide dismutase (SOD): An enzyme that is found in all living cells, where it catalyzes the destruction of O_2^- radicals.

synergistic antioxidants: Substances that form complexes with the prooxidative metal traces that are found in most fats and oils and participate in regeneration of exhausted antioxidants. The antioxidative action is not reduced until both antioxidant and synergist are completely consumed.

telomere: A region of repetitive DNA at the end of a chromosome that protects the end of the chromosome from deterioration.

testosterone: A hormone produced by the testicles and responsible for the proper development of male sexual characteristics.

tissue: Any of the distinct types of material of which animals or plants are made, consisting of specialized cells and their products.

transcription: The process in a cell by which genetic material is copied from a strand of DNA to a complementary strand of RNA (called messenger RNA).

tripinnate: Bipinnate with each division pinnate, like the leaves of some ferns.

turbid water: Water in which sediment or foreign particles is stirred up or suspended.

uterus: A female reproductive organ located between the bladder and the rectum in the pelvic area.

valence electrons octet rule: A chemical rule of thumb that reflects observation that atoms of main-group elements tend to combine in such a way that each atom has eight electrons in its valence shell, giving it the same electronic configuration as a noble gas.

vesicant: Producing a blister or blisters.

vitamin: An organic compound needed in small quantities to sustain life. The human body produces insufficient quantities of specific vitamins or none at all; therefore, we must get vitamins from our foods or from supplements.

vitamin A: A group of unsaturated nutritional organic antioxidant compounds that includes retinol, retinal, retinoic acid, several provitamin A carotenoids, and beta-carotene.

vulcanization: A chemical process for converting natural rubber or related polymers into more durable materials via the addition of sulfur or other equivalent curatives or accelerators.

Western diet: A diet common among many people in some developed countries and increasingly in developing countries. It is characterized by high intake of red meat, sugary desserts, high-fat foods, and refined grains.

xenobiotic: A chemical compound (such as a drug, pesticide, or carcinogen) that is foreign to a living organism.

xerophthalmia: Abnormal dryness of the conjunctiva and cornea of the eye with inflammation and ridge formation. It is typically associated with vitamin A deficiency.

zygote: A fertilized egg cell that results from the union of an ovum with a sperm. In the embryonic development of humans and other animals, the zygote stage is brief and is followed by cleavage, when the single cell subdivides into smaller cells. The zygote represents the first stage in the development of a genetically unique organism.

References

Aboazma, S. Biological Oxidation. PPT. http://www1.mans.edu.eg/FacMed/english/dept/biochemistry/pdf/OXIDATION.pdf.

Adesanwo, J. K., Y. Raji, S. B. Olaleye, S. A. Onasanwo, O. O. Fadare, O. O. Ige, and O. O. Odusanya. 2007. Antiulcer Activity of Methanolic Extract of *Bryophyllum Pinnatum* in Rats." Journal of Biological Sciences, 409–12.

Afzal, M., G. Gaurav, I. Kazmi, M. Rahman, O. Afzal, J. Alam, K. R. Hakeem, M. Pravez, R. Gupta, and F. Anwar. 2012. "Anti-Inflammatory and Analgesic Potential of a Novel Steroidal Derivative from *Bryophyllum pinnatum*." Fitoterapia 83: 853–858.

Afzal, M., I. Kazmi, R. Khan, R. Singh, M. Chauhan, T. Bisht, and F. Anwar. 2012. "*Bryophyllum pinnatum*: A review." International Journal of Research in Biological Sciences 2 (4): 143–149.

Agrawal, B., and A. Mehta. 2008. "Antiasthmatic Activity of *Moringa oleifera* Lam: A Clinical Study." Indian Journal of Pharmacology 40 (1): 28–31.

Akinsulire, O. R., I. E. Aibinu, T. Adenipekun, and T. Odugbemi. 2007. "In Vitro Antimicrobial Activity of Crude Extracts from Plants *Bryophyllum pinnatum* and *Kalanchoe crenata*." African Journal of Traditional Complementary Alternative Medicine 4 (3): 338–344.

Alabi, D. A., M. Z. Onibudo, and N. A. Amusa. 2005. "Chemicals and Nutritional Composition of Four Botanicals with Fungitoxic Properties." World Journal of Agricultural Sciences 1 (1): 84–88.

Alberts, B., A. Johnson, J. Lewis, M. Raff, K. Roberts, and P. Walter. 2002. "Programmed Cell Death (Apoptosis)." In Molecular Biology of the Cell. 4th ed. New York: Garland Science. http://www.ncbi.nlm.nih.gov/books/NBK26873.

Amic, Dragon, D. Davidovic-Amic, D. Beslo, and N. Trinajstic. 2003. "Structure-Radical Scavenging Activity Relationships of Flavanoids." Croatia Chemica ACTA 76 (1): 55–61. http://fulir.irb.hr/753/1/CCA_76_2003_055_061_amic.pdf.

ANC. 2015. "Cellular Health." Website of the Alternative Naturopathic Center. http://www.alternativenaturopathiccenter.com/Cellular_health.html.

Anderson, T., D. Reid, and G. Beaton. 1972. "Vitamin C and the Common Cold: A Double Blind Trial." CMA Journal 107: 503–508.

Antonio, C., P. Cristina, B. Claudia, E. Sara, S. Umberto, and M. Patrizia. 2001. "Association between Ischemic Stroke and Increased Oxidative Stress." Argentinian Federation of Cardiology. http://www.fac.org.ar/scvc/llave/stroke/cherubi/cherubini.htm.

Anwar, F., S. Latif, M. Ashraf, and A. H. Gilani. 2007. "*Moringa* oleifera: A Food Plant with Multiple Medicinal Uses." Phytotherapy Research 21: 17–25.

Atif, Ali. 2014. "Enhancement of Human Skin Facial Revitalization by Moringa Leaf Extract Cream." Postepy Dermatol Alergol 31, no. 2: 71–76.

National Institute on Aging. 2011. Biology of aging: research today for a healthier tomorrow. http://purl.fdlp.gov/GPO/gpo46777.

Beckman, K., and B. Ames. 1998. "The Free Radical Theory of Aging Matures." Physiological Reviews 78: 547–81. http://physrev.physiology.org/content/78/2/547.

Bell, E. F. 1987. "Vitamin E in Infant Nutrition." American Journal of Clinical Nutrition 46: 183–6.

Berryman, S. 2005. "Ancient Atomism." In Stanford Encyclopedia of Philosophy. Stanford University, 1997. http://plato.stanford.edu/entries/atomism-ancient/.

Bhattacharya A., M. R. Naik, D. Agrawal, K. Rath, S. Kumar, and S. S. Mishra. 2014. "Anti-Pyretic, Anti-Inflammatory and Analgesic Effects of Leaf Extract of Drumstick Tree." Journal of Young Pharmacists 6 (4): 20–24.

Blokhina, Olga, E. Virolainen, and K. Fagerstedt. 2003. "Antioxidants, Oxidative Damage and Oxygen Deprivation Stress: A Review." Annals of Botany. http%3A%2F%2Faob.oxfordjournals.org%2Fcontent%2F91%2F2%2F179.full%23sec-9.

"Cadmium Exposure and Human Health." Cadmium. Accessed November 13, 2015. http://www.cadmium.org/pg.php?id_menu=5.

Cameron E., L. Pauling, and B. Leibovitz. 1979. "Ascorbic Acid and Cancer: A Review." Cancer Research 39: 663–681.

Carpenter, K. 1988. "The History of Scurvy and Vitamin C." Cambridge, UK: Cambridge University Press.

Carr, Anitra, and B. Frei. 2010. "Toward a New Recommended Dietary Allowance for Vitamin C Based on Antioxidant and Health Effects in Humans." The American Journal of Clinical Nutrition, 1086–107.

Carrie, W. "The Older Population: 2010," United States Census 2010, November 12, 2015, http://www.census.gov/prod/cen2010/briefs/c2010br-09.pdf.

"Catalase." InterPro. https://www.ebi.ac.uk/interpro/potm/2004_9/Page2.htm.

Chambial, S., S. Dwivedi, K. K. Shukla, P. J. John, and P. Sharma. 2013. "Vitamin C in Disease Prevention and Cure: An Overview." Indian Journal of Clinical Biochemistry 28 (4): 314–328.

Chen, Q., M. G. Espey, M. C. Krishna, J. B. Mitchell, C. P. Corpe, G. R. Buettner, E. Shacter, and M. Levine. 2005. "Pharmacologic Ascorbic Acid Concentrations Selectively Kill Cancer Cells: Action as a Pro-drug to Deliver Hydrogen Peroxide to Tissues." Proceedings of the National Academy of Sciences, 13604–3609.

Chong, E. W-T, T. Y. Wong, A. J. Kreis, J. A. Simpson, and R. H. Guymer. 2007. "Dietary Antioxidants and Primary Prevention of Age Related Macular Degeneration: Systematic Review and Meta-analysis." BMJ 13: 755.

Chumark, P., P. Khunawat, Y. Sanvarinda, S. Phornchirasilp, N. P. Morales, L. Phyvthong-ngam, P. Ratanachamnong, S. Sriwawat, and K. S. Pongrapeeporn. 2008. "The In Vitro and Ex Vivo Antioxidant Properties, Hypolipidaemic and Antiatherosclerotic Activities of Water Extract of *Moringa oleifera* Lam. leaves." Journal of Ethnopharmacology.

Colagar, Abasalt H., and Eisa T. Marzony. 2009. "Ascorbic Acid in Human Seminal Plasma: Determination and Its Relationship to Sperm Quality." Journal of Clinical Biochemistry and Nutrition, 144–49.

Daniel, H. "Benefits of Oxidation," 2011. Accessed November 1, 2015. http://benefitof.net/benefits-of-oxidation/.

DeGray, A. "Seeking Immortality: Aubrey DeGray at TEDxSalford," YouTube Video, 20:09, posted by "TEDx Talks," March 25, 2014. https://www.youtube.com/watch?v=T0lvxTm2iLg.

"Development of the Atomic Theory." American Boards. Accessed November 1, 2015. http://www.abcte.org/files/previews/chemistry/s1_p5.html.

Dewiyanti, I. D., E. Filailla, and T. Y. Megawati. 2012. "The Antidiabetic Activity of Cocor Bebek Leaves (*Kalanchoe pinnata* Lam. Pers.) Ethanolic Extract from Various Areas." The Journal of Tropical Life Science 2 (2): 37–39.

Doughari, J. H., M. S. Pukuma, and N. De. 2007. "Antibacterial effects of *Balanites* aegyptiaca L. Drel. and *Moringa oleifera* Lam. on Salmonella typhi." African Journal of Biotechnology 6 (19): 2212–2215.

"The Economic Case for Health Care Reform." The White House. June 1, 2009. Accessed November 13, 2015. https://www.whitehouse.gov/administration/eop/cea/TheEconomicCaseforHealthCareReform.

Eilert, U., B. Wolters, and A. Nahrstedt. 1981. "The Antibiotic Principle of Seeds of *Moringa oleifera* and *Moringa stenopetala*." Journal of Medicinal Plant Research 42: 55–61.

Eskind Biomedical Library. 2015. "Scurvy." Accessed August 28. http://www.mc.vanderbilt.edu/diglib/sc_diglib/hc/journeys/john_woodall.html.

Estrella, M. C. P., J. B. Mantaring, G. Z. David, and M. A. Taup. 2000. "A Double-Blind, Randomized Controlled Trial on the Use of Malunggay (*Moringa oleifera*) for Augmentation Volume of Breastmilk among Non-Nursing Mothers of Preterm Infants." The Philippine Journal of Pediatrics 49 (1): 3–6.

"Facts About Sarin." CDC. 2013. Accessed November 12, 2015. http://www.bt.cdc.gov/agent/sarin/basics/facts.asp.

Fell, D. A., and S. Thomas. 1995. "Physiological Control of Metabolic Flux: The Requirement of Multisite Modulation." Biochemistry Journal 311 (1): 35–39.

Ferreira, I., Q. Maria-Joao, B. Miguel, E. Letecia, B. Agathe, and K. Gilbert. 2006. "Evaluation of Antioxidant Properties." Research Gate. http://www.researchgate.net/publication/7444593_Evaluation_of_the_antioxidant_properties_of_diarylamines_in_the_benzobthiophene_series_by_free_radical_scavenging_activity_and_reducing_power.

Flora, S. 2009. "Structural, Chemical and Biological Aspects of Antioxidants for Strategies against Metal and Metalloid Exposure." Oxidative Medicine and Cellular Longevity. Landes Bioscience. http://www.ncbi.nlm.nih.gov/pmc/articles/PMC2763257/.

"Food Sources of Vitamin C." The Dietitians of Canada. http://www.dietitians.ca/Your-Health/Nutrition-A-Z/Vitamins/Food-Sources-of-Vitamin-C.aspx.

Garnham, C. 2006. "Two Halves of a Redox Equation." https://commons.wikimedia.org/wiki/File:Redox_Halves.png.

Smart, Gene. 2015. "High Polyphenols Foods and Polyphenols Sources." Gene Smart Wellness. https://www.genesmart.com/pages/high_polyphenols_foods/161.php.

"Genetics." In Biology of Aging. National Institute of Aging, 2011.

Ghasi, S., E. Nwobodo, and J. O. Ofili. 2000. "Hypocholesterolemic Effects of Crude Extract of Leaf of *Moringa oleifera* Lam. in High-Fat Diet Fed Wistar Rats." Journal of Ethnopharmacology 69: 21–25.

Ghasi, S., C. Egwuibe, P. U. Achukwu, and J. C. Onyeanusi. 2011. "Assessment of the Medical Benefit in the Folkloric Use of *Bryophyllum pinnatum* Leaf among the Igbos of Nigeria for the Treatment of Hypertension." *African Journal of Pharmacy and Pharmacology* 5 (1): 83–92.

Giridhari, V. V. A., D. Malathi, and K. Geetha. 2011. "Antidiabetic Property of Drumstick (*Moringa oleifera*) Leaf Tablets." International Journal of Health and Nutrition 2 (1): 1–5.

"The Global Burden." In IDF Diabetes Atlas, 2014. 29–48. 6th ed. International Diabetes Foundation, 2014.

Goebel, L., J. Wong, and V. Perry. 2013. "Scurvy." Medscape. http://emedicine.medscape.com/article/125350-overview.

Gospodaryov, D., and L. Volodymyr. 2012. "Oxidative Stress: Cause and Consequence of Diseases." INTECH Open Access. http://cdn.intechopen.com/pdfs/35941/InTech-Oxidative_stress_cause_and_consequence_of_diseases.pdf.

Guevara, A. P., C. Vargas, H. Sakurai, Y. Fujiwara, K. Hashimoto, T. Maoko, M. Kozuka, Y. Ito, H. Tokuda, and H. Nishino. 1999. "An Antitumor Promoter from *Moringa oleifera* Lam." Mutation Research 440: 181–188.

Gwehenberger, B., C. Rist, R. Huch, and V. von Mandach. 2004. "Effects of *Bryophyllum pinnatum* versus Fenoterol on Uterine Contractility." European Journal of Obstetrics, Gynaecology and Reproductive Biology 113: 164–171.

Halberstein, R. A. 2005. "Medicinal Plants: Historical and Cross-Cultural Usage Patterns." Ann Epidemiology 15 (9): 686–699.

Hammed, M. 2012. "Quantum Numbers." Lecture presented November 12. http://www.uobabylon.edu.iq/eprints/publication_12_21320_24.pdf.

Hans, U., and B. Edward. 2010. "Regular Vitamin C Supplementation during Pregnancy Reduces Hospitalization: Outcomes of a Ugandan Rural Cohort Study." Pan African Medical Journal 5 (5).

Harman, D. 1956. "Aging: A Theory Based on Free Radical and Radiation Chemistry." Journal of Gerontology, 298–300.

Harris, E. 1992. "Regulation of Antioxidant Enzymes." Federation of American Societies for Experimental Biology 6: 2675–2683. http://www.fasebj.org/content/6/9/2675.full.pdf.

Harris, I. 2011. Healing Herbs of Jamaica. AhHa Press.

Harrison, F. E., G. L. Bowman, and M. C. Polidori. 2014. "Ascorbic Acid and the Brain: Rationale for the Use against Cognitive Decline." Nutrients 6: 1752–1781.

Hart, Y. 2015. "Support Your Cellular Health and Slow the Aging Process." http://yolandehart.com/support-your-cellular-health-and-slow-the-aging-process/.

Heitzer T., H. Just, and T. Münzel. 1996. "Antioxidant Vitamin C Improves Endothelial Dysfunction in Chronic Smokers." Circulation 94: 6–9.

Helmenstine, A. M. 2014. "Elemental Composition of the Human Body." http://chemistry.about.com/od/biochemistry/tp/Chemical-Composition-Of-The-Human-Body.htm.

Herzog, G. 2014. "Isotope." Encyclopedia Britannica Online. http://www.britannica.com/science/isotope.

Hill, M. F., and P. K. Singal. 1996. "Antioxidant and Oxidative Stress Changes during Heart Failure Subsequent to Myocardial Infarction in Rats." American Journal of Pathology 148, no. 1: 291–300.

Hoffer, A. 1989. "The Discovery of Vitamin C Albert Szent-Györgyi, M.D., Ph.D., 1893–1986." Journal of Orthomolecular Medicine 4 (1): 24–26. http://www.orthomolecular.org/library/jom/1989/pdf/1989-v04n01-p024.pdf.

Howard, L., A. Wong, A. Perry, and B. Klein. 1999. "Beta-Carotene and Ascorbic Acid Retention in Fresh Food and Processed Vegetables." Journal of Food Science 64: 929–956.

Igwemmar, N. C., S. A. Kolawole, and I. A. Imran. 2013. "Effect of Heating on Vitamin C Content of Some Selected Vegetables." International Journal of Scientific & Technology Research 2 (11): 209–212.

"Introduction to Particle Physics." THE BIRTH OF THE ATOM. 1999. Accessed November 1, 2015. http://molaire1.perso.sfr.fr/e_histoire.html.

"Isotopes." University of Colorado. http://www.colorado.edu/physics/2000/isotopes/stable_isotopes.html.

James. 2013. "In Summary: Free Radicals." Master Organic Chemistry. http://www.masterorganicchemistry.com/2013/12/09/in-summary-free-radicals/.

Jensen, W. 2007. "The Origin of the Oxidation-State Concept." Journal of Chemical Education 84 (9): 1418. http://www.che.uc.edu/jensen/W.%20B.%20Jensen/Reprints/139.%20Oxidation%20States.pd.

Berg, J. M. Biochemistry, 5th ed. W.H. Freeman, 2002.

Johnston, C. S., G. M. Barkyoumb, and S. S. Schumacher. 2014. "Vitamin C Supplementation Slightly Improves Physical Activity Levels and Reduces Cold Incidence in Men with Marginal Vitamin C Status: A Randomized Controlled Trial." Nutrients 6: 2572–2583.

Jung, I. L. 2014. "Soluble Extract from *Moringa oleifera* Leaves with a New Anticancer Activity." PLoS One 9 (4).

Kaarteenaho-Wiik, R., and V. Kinnula. 2004. "Distribution of Antioxidant Enzymes in Developing Human Lung, Respiratory Distress Syndrome, and Bronchopulmonary Dysplasia." Journal of Histochemistry and Cytochemistry 52 (9): 1231–240.

Kallner, A., D. Hartmann, and D. Hornig. 1981. "On the Requirement of Ascorbic Acid in Man: Steady-state Turnover and Body Pool in Smokers." American Journal of Clinical Nutrition 34: 1347–355.

Kamboj, A., and A. K. Saluja. 2009. "*Bryophyllum pinnatum* (Lam.) Kurz. Phytochemical and Pharmacological Profile: A Review." Pharmocognosy Reviews 3: 364–374.

Kaur, N., R. Bains, and J. Niazi. 2014. "A Review on *Bryophyllum pinnatum*—A Medicinal Herb." Journal of Medical and Pharmaceutical Innovation 1 (3): 13–19.

Kim Woo, J., M. Jo, S. Jung, B. Jee, D. Choi, and Q. Jo. 2002. "Combined Effects of Copper and Temperature on Antioxidant Enzymes in the Black Rockfish Sebastes schlegeli." Fisheries and Aquatic Sciences 5 (3): 200–205.

Kitani, K. 2007. "What Really Declines with Age?" Age 29 (1): 1–14. National Center for Biotechnology Information.

Lals, S., and J. Tsaknis. 2002. "Characterization of *Moringa oleifera* Seed Oil Variety 'Periyakulam 1.'" Journal of Food Composition and Analysis 15: 65–77.

Lans, C. A. 2006. "Ethnomedicines Used in Trinidad and Tobago for Urinary Problems and Diabetes Mellitus." Journal of Ethnobiology and Ethnomedicine 2 (45).

"Leucippus and Democritus." 2004. Purdue University. http://chemed.chem.purdue.edu/genchem/history/atom.html.

Levine, G. L., B. Frei, S. N. Koulouris, M. D. Gerhard, J. F. Keaney, and J. A. Vita. 1996. "Ascorbic Acid Reverses Endothelial Vasomotor Dysfunction in Patients with Coronary Artery Disease." Circulation 93: 1107–1113.

"Lewis Formulas and Octet Rule." http://www.chemie-biologie.unisiegen.de/ac/hjd/lehre/advanced_vortraege0607/chen_lewis_corr.pdf.

"Lipoic Acid." 2007. Life Extension. http://www.lifeextension.com/magazine/2007/10/nu_lipoic_acid/page-01.

Lobo, V., A. Patil, A. Phatak, and N. Chandra. 2010. "Free Radicals, Antioxidants and Functional Foods: Impact on Human Health." Pharmacognosy Reviews 4 (8): 118–126. http://www.ncbi.nlm.nih.gov/pmc/articles/PMC3249911/.

Mahata, S., S. Maru, S. Shukla, A. Pandey, G. Mugesh, B. C. Das, and A. C. Bharti. 2012. "Anticancer Property of Bryophyllum pinnata (Lam.) Oken. Leaf on Human Cervical Cancer Cells." BioMed Central Complementary and Alternative Medicine 12: 15.

Mahmood, K. T., T. Mugal, and I. Ul Haq. 2010. "*Moringa oleifera*: A Natural Gift—A Review." Journal of Pharmaceutical Sciences & Research 2 (11): 775–781.

Mahmood, K. T., T. Mugal, and I. Ul Haq. "*Moringa Oleifera*: A Natural Gift—A Review." Journal of Pharmaceutical Science and Research 2, no. 11 (2010): 775–81.

Mandal, A. 2010. "What Is Oxidative Stress?" News-Medical.net. http://www.news-medical.net/health/What-is-Oxidative-Stress.aspx.

Mbikay, M. 2012. "Therapeutic Potential of *Moringa oleifera* Leaves in Chronic Hyperglycemia and Dyslipidemia: A Review." Frontiers in Pharmacology 3 (24).

The Merck Manual: Home Edition. "The Aging Body." http://www.msdmanuals.com/home/older-people-s-health-issues/the-aging-body/changes-in-the-body-with-aging.

Mercola. "The Ultimate Guide to Antioxidants." http://articles.mercola.com/antioxidants.aspx.

M. oleifera. PubChem database. www.pubmed.ncbi.nlm.nih.org/.

Mudi, S. Y., and H. Ibrahim. 2008. "Activity of *Bryophyllum pinnatum* S. Kurz Extracts on Respiratory Tract Pathogenic Bacteria." Bayero Journal of Pure and Applied Sciences 1 (1): 43–48.

Naidu, K. A. 2003. "Vitamin C in Human Health and Disease Is Still a Mystery? An Overview." Nutrition Journal 2: 7.

National Council on Aging. 2012. "A Look at How the Environment Impacts Healthy Aging." Website of the National Council on Aging. http://www.ncoa.org/improve-health/chronic-conditions/a-look-at-how-the-environment.html.

National Institute on Aging. 2008. Healthy aging lessons from the Baltimore Longitudinal Study of Aging. [Bethesda, Md.]: National Institute on Aging, National Institutes of Health, US Dept. of Health and Human Services. http://purl.access.gpo.gov/GPO/LPS115258.

NIH. 2013. "Vitamin A." Health professional fact sheet. National Institutes of Health. https://ods.od.nih.gov/factsheets/VitaminA-HealthProfessional/#h5.

NIH. 2013. "Vitamin C." Consumer Fact Sheet. National Institutes of Health. https://ods.od.nih.gov/factsheets/VitaminC-Consumer/.

NIH. 2013. "Vitamin C." Health professional fact sheet. National Institutes of Health. https://ods.od.nih.gov/factsheets/VitaminC-HealthProfessional/.

NIH. 2013. "Vitamin E." Health professional fact sheet. National Institutes of Health. https://ods.od.nih.gov/factsheets/VitaminE-HealthProfessional/.

NIH. 2013. "Selenium." Health professional fact sheet. National Institutes of Health. https://ods.od.nih.gov/factsheets/Selenium-HealthProfessional/.

Nogales. "Signaling 1." 5 CELL SIGNALING I Introduction. 2008. Accessed November 1, 2015. https://mcb.berkeley.edu/courses/mcb110spring/nogales/mcb110_s2008_4signaling.pdf.

Norman, R. 2014. "Chemical Compound." Encyclopedia Britannica Online. http://www.britannica.com/EBchecked/topic/108614/chemical-compound.

Nourished Media. 2009. "Fat Soluble Vitamins: Vitamins A, D, E & K." Nourished Kitchen. Nourished Media. http://nourishedkitchen.com/fat-soluble-vitamins/.

Nwali, B. U., A. N. C. Okaka, C. E. Offor, P. M. Aja, and U. E. Nwachi. 2014. "Proximate and Mineral Compositions of *Bryophyllum Pinnatum* Leaves." American Journal of Phytomedicine and Clinical Therapeutics 2, no. 3: 286–89.

Ogbonnia, S. O., J. I. Odimegwu, and V. N. Enwuru. 2008. "Evaluation of Hypoglycaemic and Hypolipidaemic Effects of Aqueous Ethanolic Extracts of Treculia africana Decne and *Bryophyllum pinnatum* Lam. and Their Mixture on Streptozotocin (STZ)-Induced Diabetic Rats." African Journal of Biotechnology 7 (15): 2535–2539.

Okwu, D. E., and C. Josiah. 2006. "Evaluation of the Chemical Composition of Two Nigerian Medicinal Plants." African Journal of Biotechnology 5 (4): 357–361.

Okwu, D., and F. Nnamdi. 2011. "Two Novel Flavonoids from *Bryophyllum Pinnatum* and Their Antimicrobial Activity." Journal of Chemistry and Pharmaceutical Research 3, no. 2: 1–10.

Okwu, D., and F. Nnamdi. "A Novel Antimicrobial Phenanthrene Alkaloid from *Bryopyllum Pinnatum*." E-Journal of Chemistry 8, no. 3 (2009): 1456–461.

O'Niel, P. 1971. "The Vitamin C Mania." Life July 9: 55–62.

Ozolua, R. I., C. J. Eboka, C. N. Duru, and D. O. Uwaya. 2010. "Effects of Aqueous Leaf Extract of *Bryophyllum pinnatum* on Guinea Pig Tracheal Ring Contractility." Nigeria Journal of Physiological Science 25: 149–157.

Ozolua, R. I., S. E. Idogun, and G. E. Tafamel. 2010. "Acute and Sub-Acute Toxicological Assessment of Aqueous Leaf Extract of *Bryophyllum Pinnatum* (Lam.) in Sprague-Dawley Rats." American Journal of Pharmacology and Toxicology 5 (3): 145–151.

Pal, S., and A. K. Chaudhari. 1991. "Studies on the Anti-Ulcer Activity of a *Bryophyllum Pinnatum* Leaf Extract in Experimental Animals." Journal of Ethnopharmacology 33: 97–102.

Pauling, L. 1971. "The Significance of the Evidence about Ascorbic Acid and the Common Cold." Proceedings of the National Academy of Sciences of the USA 68 (11): 2678–2681.

Pauling, L. 1971. "Vitamin C and the Common Cold." CMA Journal 105: 448–449.

Pearson Education Inc. "A Closer Look at Electron Carriers." http://www.phschool.com/science/biology_place/biocoach/cellresp/closer2.html.

Percival, M. 1998. "Antioxidants." In Clinical Nutrition Insights: 1–4. Advanced Nutrition Publications Inc. http://acudoc.com/Antioxidants.pdf.

Perez-Campo, R., M. Lopes-Torrez, S. Cadenas, C. Rojas, and G. Barja. 1998. "The Rate of Free Radical Production as a Determinant of the Rate of Aging: Evidence from the Comparative Approach." Journal of Comparative Physiology 149–58. http://www.life.umd.edu/faculty/wilkinson/honr278c/PDF/Perez-Campo98.pdf.

Peroni, L., R. Ferreira, A. Figuiera, M. Machado, and D. Stach-Machado. 2007. "Expression Profile of Oxidative and Antioxidative Stress Enzymes Based on ESTs Approach of Citrus." SciElo. http://www.scielo.br/scielo.php?pid=S1415-47572007000500016&script=sci_arttext.

Plangger, N., L. Rist, R. Zimmerman, and U. Mandach. 2006. "Intravenous Tocolysis with *Bryophyllum pinnatum* Is Better Tolerated than Beta-Agonist Application." European Journal of Obstetrics & Gynecology and Reproductive Biology 124: 168–172.

Powell, S. 2000. "The Antioxidant Properties of Zinc." Journal of Nutrition. http://jn.nutrition.org/content/130/5/1447S.full.pdf+html.

Prabhu, K., K. Murugan, A. Nareshkumar, N. Ramasubramanian, and S. Bragadeeswaran. 2011. "Larvicidal and Repellent Potential of *Moringa oleifera* against Malarial Vector, Anopheles stephensi Liston (Insecta: Diptera: Culicidae)." Asian Pacific Journal of Tropical Biomedicine 1 (2): 124–129.

Prabsattroo, T., J. Wattanathorn, S. Iamsaard, P. Somsapt, O. Sritragool, W. Thokhummee, and S. Muchimapura. 2015. "*Moringa oleifera* Extract Enhances Sexual Performance in Stressed Rats." Journal of Zhejiang University Science B 16 (3): 179–190.

Price, C. 2015. The Vitamin C Complex: Our Obsessive for Nutritional Perfection. One World Publications.

Price, M. L. (1985, 2000, 2002) 2007. "The Moringa Tree." Echo Technical Note.

"Proton." Hyper Physics. Georgia State University. http://hyperphysics.phy-astr.gsu.edu/hbase/particles/proton.html.

Radford, D. J., A. D. Gillies, and J. A. Hinds. 1986. "Naturally Accruing Cardiac Glycoside." Medical Journal of Australia 144: 540–44.

Raj, A., M. P. Gururaja, H. Joshi, and C. S. Shastry. 2014. "Kalanchoe pinnatum in Treatment of Gallstones: An Ethnopharmacological Review." Internal Journal of PharmTech Research 6 (1): 252–261.

Rastogi, T., V. Bhutda, K. Moon, P. B. Aswar, and S. S. Khadabadi. 2009. "Comparative Studies on Anthelmintic Activity of *Moringa oleifera* and Vitex negundo." Asian Journal of Research Chemistry 2 (2): 181–182.

Reuter, S., S. Gupta, M. Chaturvedi, and B. Aggarwal. 2010. "Oxidative Stress, Inflammation, and Cancer: How Are They Linked?" Free Radical Biology & Medicine. Accessed November 1, 2015. http://www.ncbi.nlm.nih.gov/pmc/articles/PMC2990475/.

Riddle, D. L., T. Blumenthal, B. J. Meyer, et al., eds. 1997. C. Elegans II. 2nd ed. Cold Spring Harbor, NY: Cold Spring Harbor Laboratory Press.

Romao, S. "Therapeutic Value of Oral Supplementation with Melon Superoxide Dismutase and Wheat Gliadin Combination." Nutrition, 2015, 430–36.

Rosenfield, L. 1997. "Vitamine–Vitamin: The Early Years of Discovery." Clinical Chemistry 43 (4): 680–685. http://www.clinchem.org/content/43/4/680.long.

Sahiner, U., B. Esra, E. Serpil, S. Cansin, and K. Omer. 2011 "Oxidative Stress in Asthma." The World Allergy Organization Journal. http://www.ncbi.nlm.nih.gov/pmc/articles/PMC3488912/.

Salahdeen, H. M., and O. K. Yemitan. 2006. "Neuropharmacological Effects of Aqueous Leaf Extract of *Bryophyllum pinnatum* in Mice." African Journal of Biomedical Research 9: 101–107.

Sanghera, P. 2011. "Quantum Physics for Scientists and Technologists." Google Books. http://www.ias.ac.in/resonance/Volumes/15/01/0016-0031.pdf.

Santariano, W. 2005. Aging, Health, and the Environment: An Ecological Model. Jones and Bartlett Learning. 1–40. http://www.jblearning.com/samples/0763726559/SampleChapter02.pdf.

Sarsour, E., M. Kumar, L. Chaudhuri, A. Kalen, and P. Goswami. 2009. "Redox Control of the Cell Cycle in Health and Disease." Antioxidants & Redox Signaling. Mary Ann Liebert Inc. http://www.ncbi.nlm.nih.gov/pmc/articles/PMC2783918/.

Satish, A., S. Sairam, F. Ahmed, and A. Urug. 2012. "*Moringa oleifera* Lam.: Protease Activity Against Blood Coagulation Cascade." Pharmacognosy Research 4 (1): 44–49.

Schultz, J. 2015. "Albert Szent-Györgyi's Discovery of Vitamin C." Accessed August 28. http://www.acs.org/content/acs/en/education/whatischemistry/landmarks/szentgyorgyi.html#discovery-of-ascorbic-acid.

Semba. 2012. "On the Discovery of Vitamin A." National Center for Biotechnology Information. US National Library of Medicine. http://www.ncbi.nlm.C.gov/pubmed/23183288.

Sethi N., D. Nath, S. C. Shukla, and R. Dyal. 1988. "Abortifacient Activity of a Medicinal Plant *Moringa oleifera* in Rats." Anc Sci Life 7 (3–4): 172–174.

Sharma, P., P. Kumari, M. M. Srivastava, and S. Srivastava. 2006. "Removal of Cadmium from Aqueous System by Shelled *Moringa oleifera* Lam. Seed Powder." Bioresource Technology 97 (2): 299–305.

Sharma, U., M. Lahkar, and J. Lahon. 2012. "Evaluation of Antidiarrhoeal Potential of *Bryophyllum pinnatum* in Experimental Animals." Asian Journal of Biomedical and Pharmaceutical Sciences 2 (15): 28–31.

Sherman, B., C. Clark, and H. Sherman. 1998. "Free Radicals: Effects, Causes, and Defenses." Personal website. http://bsherman.net/freeradicals.htm.

Sigma-Aldrich. 2007. "Dietary Antioxidants." Sigma-Aldrich website. http://www.sigmaaldrich.com/technical-documents/articles/biofiles/dietary-antioxidants0.html.

Sinatra, S. 2015. "Nitric Oxide Benefits Cardiovascular Health." Personal website. http://www.drsinatra.com/nitric-oxide-benefits-cardiovascular-health.

Soyer, O.S., M. Salathe, and S. Bonhoeffer. 2006. "Signal Transduction Networks: Topology, Response, and Biological Processes." Journal of Theoretical Biololgy 238 (2): 416–425.

Spargias, K., E. Alexopoulos, S. Kyrzopoulos, P. Iacovis, D. C. Greenwood, A. Manginas, V. Voudris, G. Pavlides, C. E. Buller, D. Kremastinos, and D. V. Cokkinos. 2004. "Ascorbic Acid Prevents Contrast-Mediated Nephropathy in Patients with Renal Dysfunction Undergoing Coronary Angiography or Intervention." Circulation 110: 2837–2842.

Steffan, J. 2012. "Tadeusz Reichstein: The Alchemist of Vitamin Science." Sight and Life 44–48.

Straaten, H. M. O., A. M. Spoelstra-de Man, and M. C. de Waard. 2014. "Vitamin C revisited." Critical Care 18: 460.

Strassler, M. 2013. "Protons and Neutrons: The Massive Pandemonium in Matter." Of Particular Significance. http://profmattstrassler.com/articles-and-posts/particle-physics-basics/the-structure-of-matter/protons-and-neutrons/.

Stubbs, B. 2003. "Captain Cook's Beer: The Antiscorbutic Use of Malt and Beer in Late 18th Century Sea Voyages." Asia Pacific Journal of Clinical Nutrition 12 (2): 129–137.

Stur, M., A. Reitner, M. Tittl, and V. Meisinger. 1996. "Oral Zinc and the Second Eye in Age-related Macular Degeneration." Invest Ophthamology Visual Science 37, no. 7, 1225-235.

Suckow, B., and M. Suckow. 2006. "Lifespan Extension by the Antioxidant Curcumin in Drosophila Melanogaster." International Journal of Biomedical Science. http://www.ncbi.nlm.nih.gov/pmc/articles/PMC3614642/.

Summers, V. 2014. "Important Differences between Bound and Unbound Electrons." In Decoded Science. http://www.decodedscience.com/important-differences-bound-unbound-electrons/50620.

Sutherland, J., G. Folkard, M. A. Mtawli, and W. D. Grant. 1994. "*Moringa Oleifera* as a Natural Coagulant," 297–99.

Sutton, C. "Antimatter." Encyclopedia Britannica Online. http://www.britannica.com/EBchecked/topic/28179/antimatter.

Sutton, C. 2014. "Subatomic Particle." Encyclopedia Britannica Online. http://www.britannica.com/science/subatomic-particle.

Szeto, Y. T., B. Tomlinson, and F. Benzie. 2002. "Total Antioxidant and Ascorbic Acid Content of Fresh Fruits and Vegetables: Implications for Dietary Planning and Food Preservation." British Journal of Nutrition 87, no. 1: 55–9.

Tabussum, W., A. Kullu, and M. P. Sinha. "Effects of Leaf Extracts of *Moringa oleifera* on Regulation of Hypothyroidism and Lipid Profile." The Biascan 8, no. 2 (2013): 665–69.

Tan, D., L. Manchester, R. Reiter, W. Qi, M. Karbownik, and J. Calvo. 2000. "Significance of Melatonin in Antioxidative Defense System: Reactions and Products." Biol Signals Recept Neurosignals 9 (3–4): 137–159.

Tatsimo, S. J. N., J. D. Tamokou, L. Havyarimana, D. Csupor, P. Forgo, J. Hohmann, J. Kuiate, and P. Tane. 2012. "Antimicrobial and Antioxidant Activity of Kaempferol Rhamnoside Derivatives from *Bryophyllum pinnatum*." Biomed Central Research Notes 5: 158.

Therapeutic Research Center. 2015. "Zinc." MedlinePlus. Therapeutic Research Center. https%3A%2F%2Fwww.nlm.nih.gov%2Fmedlineplus%2Fdruginfo%2Fnatural%2F982.html.

Tinggi, U. 2008. "Selenium: It's Role as Antioxidant in Human Health." Environmental Health and Preventive Medicine. http://www.ncbi.nlm.nih.gov/pmc/articles/PMC2698273/.

Trachootham, D., L. Weiqin, M. Ogasawara, N. Rivera-Del, and P. Huang. 2008. "Redox Regulation of Cell Survival." Antioxidants & Redox Signaling. Mary Ann Liebert, Inc. http://www.ncbi.nlm.nih.gov/pmc/articles/PMC2932530/.

Trefil, J. 2015. "Quantum Field Theory and the Standard Model." Encyclopedia Britannica Online. http://www.britannica.com/science/atom.

Tröhler, U. 2003. "James Lind and Scurvy: 1747 to 1795." JLL Bulletin. http://www.jameslindlibrary.org/articles/james-lind-and-scurvy-1747-to-1795/.

Tsutsui, H. 2005. "Mitochondrial Oxidative Stress and Heart Failure -Novel Pathophysiological Insight and Treatment Strategies." Current Cardiology Reviews, 37–44.

Tsutsui, H., K. Shintaro, and M. Shouji. 2011. "Oxidative Stress and Heart Failure." American Physiological Society. http://ajpheart.physiology.org/content/301/6/H2181.

Turrens, J. 2004. "Mitochondrial Formation of Reactive Oxygen Species." The Journal of Physiology 552 (2). http://onlinelibrary.wiley.com/doi/10.1111/j.1469-7793.2003.00335.x/pdf.

UCSB Science Line. 2015. "Why do apples turn brown when you cut them? Why does it help if you put them in the refrigerator?" National Science Foundation. http://scienceline.ucsb.edu/getkey.php?key=64.

Ufelle, S. A., E. O. Ukaejiofo, E. E. Neboh, P. U. Achukwu, U. I. Nwagha, and S. Ghasi. 2011. "The Effect of Crude Methanolic Leaf Extract of *Bryophyllum pinnatum* on Some Haematological Parameters in Wistar Rats." Research Journal of Pharmacology 5 (2): 14–17.

Venturi, M., and S. Venturi. 2007. "Evolution of Dietary Antioxidants." European EpiMarker 11 (3): 2–8. http://www.researchgate.net/profile/Sebastiano_Venturi2/publication/234162439_epimarker_3_07_Antioxidants/links/02bfe50fa320b02e18000000.pdf.

Venugopalan, A. 2010. "Quantum Interference of Molecules Probing the Wave of Nature of Matter." Resonance. http://www.ias.as.in/resonance/Volumes/15/01/0016-0031.pdf.

Vimala, G., and F. G. Shoba. 2014. "A Review on Antiulcer Activity of Few Indian Medicinal Plants." International Journal of Microbiology, 519–590.

Wanasundara, S. 2015. "Antioxidants: Science, Technology, and Applications." Bailey's Industrial Oil and Fat Products. http://onlinelibrary.wiley.com/doi/10.1002/047167849X.bio002/full.

Watanabe, F., E. Hashizume, G. Chan, and A. Kamimura. 2014. "Skin-Whitening and Skin-Condition-Improving Effects of Topical Oxidized Glutathione: A Double-Blind and Placebo-Controlled Clinical Trial in Healthy Women." National Center for Biotechnical Information. Pub Med Central. http%3A%2F%2Fwww.ncbi.nlm.nih.gov%2Fpmc%2Farticles%2FPMC4207440%2F.

Watt, D. F. 2014. Geriatric Neurology. 1st ed. John Wiley & Sons.

Website of Trees for Life International. www.treesforlife.org/sites/default/files/documents/English%20moringa_book_view.pdf.

Webster, Inc. Merriam-Webster's Collegiate Encyclopedia. Springfield, Mass.: Merriam-Webster, 2000.

"What Exactly Is the 'Spin' of Subatomic Particles Such as Electrons and Protons? Does It Have Any Physical Significance, Analogous to the Spin of a Planet?" 1999. Scientific American Global. http://www.scientificamerican.com/article/what-exactly-is-the-spin/.

Whitton, B. 2000. Ecology of Cyanobacteria II: Their Diversity in SpaceTime and Time. Springer Science and Business Media.

Williams, M. 2014. "John Dalton's Atomic Model." Universe Today. http://www.universetoday.com/38169/john-daltons-atomic-model/.

Young Min, Eun, Su Kyong Baeck, and Ju-Chan Kang. 2014. "Combined Effects of Copper and Temperature on Antioxidant

Enzymes in the Black Rockfish Sebastes Schlegeli." Fish Aquatic Science 17, no. 3: 345–53.

Yoruk, Ruhiye, and Marshall. 2003. "Physicochemical Properties and Function of Plant Polyphenol Oxidase: A Review." http://nfscfaculty.tamu.edu/talcott/courses/FSTC605/Papers%20Reviewed/Review-PPO.pdf.

Index

B

C

D

E

J

K

L

M

N

O

P

Q

R

S

T

U

V

Printed in the United States
By Bookmasters